PROPERTY OF UNITED STATES

COMBINATORIAL DATA ANALYSIS

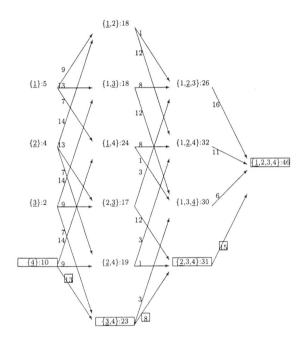

SIAM Monographs on Discrete Mathematics and Applications

The series includes advanced monographs reporting on the most recent theoretical, computational, or applied developments in the field; introductory volumes aimed at mathematicians and other mathematically motivated readers interested in understanding certain areas of pure or applied combinatorics; and graduate textbooks. The volumes are devoted to various areas of discrete mathematics and its applications.

Mathematicians, computer scientists, operations researchers, computationally oriented natural and social scientists, engineers, medical researchers, and other practitioners will find the volumes of interest.

Editor-in-Chief
Peter L. Hammer, RUTCOR, Rutgers, The State University of New Jersey

Editorial Board

M. Aigner, *Freie Universität Berlin, Germany*
N. Alon, *Tel Aviv University, Israel*
E. Balas, *Carnegie Mellon University, USA*
C. Berge, *E. R. Combinatoire, France*
J.- C. Bermond, *Université de Nice–Sophia Antipolis, France*
J. Berstel, *Université Marne-la-Vallée, France*
N. L. Biggs, *The London School of Economics, United Kingdom*
B. Bollobás, *University of Memphis, USA*
R. E. Burkard, *Technische Universität Graz, Austria*
D. G. Corneil, *University of Toronto, Canada*
I. Gessel, *Brandeis University, USA*
F. Glover, *University of Colorado, USA*
M. C. Golumbic, *Bar-Ilan University, Israel*
R. L. Graham, *AT&T Research, USA*
A. J. Hoffman, *IBM T. J. Watson Research Center, USA*
T. Ibaraki, *Kyoto University, Japan*
H. Imai, *University of Tokyo, Japan*
M. Karoński, *Adam Mickiewicz University, Poland, and Emory University, USA*
R. M. Karp, *University of Washington, USA*
V. Klee, *University of Washington, USA*
K. M. Koh, *National University of Singapore, Republic of Singapore*
B. Korte, *Universität Bonn, Germany*

A. V. Kostochka, *Siberian Branch of the Russian Academy of Sciences, Russia*
F. T. Leighton, *Massachusetts Institute of Technology, USA*
T. Lengauer, *Gesellschaft für Mathematik und Datenverarbeitung mbH, Germany*
S. Martello, *DEIS University of Bologna, Italy*
M. Minoux, *Université Pierre et Marie Curie, France*
R. Möhring, *Technische Universität Berlin, Germany*
C. L. Monma, *Bellcore, USA*
J. Nešetřil, *Charles University, Czech Republic*
W. R. Pulleyblank, *IBM T. J. Watson Research Center, USA*
A. Recski, *Technical University of Budapest, Hungary*
C. C. Ribeiro, *Catholic University of Rio de Janeiro, Brazil*
H. Sachs, *Technische Universität Ilmenau, Germany*
A. Schrijver, *CWI, The Netherlands*
R. Shamir, *Tel Aviv University, Israel*
N. J. A. Sloane, *AT&T Research, USA*
W. T. Trotter, *Arizona State University, USA*
D. J. A. Welsh, *University of Oxford, United Kingdom*
D. de Werra, *École Polytechnique Fédérale de Lausanne, Switzerland*
P. M. Winkler, *Bell Labs, Lucent Technologies, USA*
Yue Minyi, *Academia Sinica, People's Republic of China*

Series Volumes

Hubert, L., Arabie, P., and Meulman, J., *Combinatorial Data Analysis: Optimization by Dynamic Programming*
Peleg, D., *Distributed Computing: A Locality-Sensitive Approach*
Wegener, I., *Branching Programs and Binary Decision Diagrams: Theory and Applications*
Brandstädt, A., Le, V. B., and Spinrad, J. P., *Graph Classes: A Survey*
McKee, T. A. and McMorris, F. R., *Topics in Intersection Graph Theory*
Grilli di Cortona, P., Manzi, C., Pennisi, A., Ricca, F., and Simeone, B., *Evaluation and Optimization of Electoral Systems*

COMBINATORIAL DATA ANALYSIS
Optimization by Dynamic Programming

Lawrence Hubert
University of Illinois Urbana-Champaign
Champaign, Illinois

Phipps Arabie
Rutgers University
Newark, New Jersey

Jacqueline Meulman
Leiden University
Leiden, The Netherlands

siam

Society for Industrial and Applied Mathematics
Philadelphia

Copyright ©2001 by the Society for Industrial and Applied Mathematics.

10 9 8 7 6 5 4 3 2 1

All rights reserved. Printed in the United States of America. No part of this book may be reproduced, stored, or transmitted in any manner without the written permission of the publisher. For information, write to the Society for Industrial and Applied Mathematics, 3600 University City Science Center, Philadelphia, PA 19104-2688.

Library of Congress Cataloging-in-Publication Data
Hubert, Lawrence J.
    Combinatorial data analysis : optimization by dynamic programming / Lawrence Hubert, Phipps Arabie, Jacqueline Meulman.
      p. cm. – (SIAM monographs on discrete mathematics and applications)
    Includes bibliographical references and index.
    ISBN 0-89871-478-8
      1. Dynamic programming. 2. Combinatorial optimization. I. Arabie, Phipps.
  II. Meulman, Jacqueline. III. Title. IV. Series.

T57.83 .H83 2001
519.7´03--dc21

00-049656

siam is a registered trademark.

To J. Douglas Carroll
fabri meliori rationum viarumque

Contents

Preface ix

1 **Introduction** 1

2 **General Dynamic Programming Paradigm** 7
   2.1 An Introductory Example: Linear Assignment 7
   2.2 The GDPP 13

3 **Cluster Analysis** 17
   3.1 Partitioning 17
      3.1.1 Admissibility Restrictions on Partitions 21
      3.1.2 Partitioning Based on Two-Mode Proximity
           Matrices 25
   3.2 Hierarchical Clustering 28
      3.2.1 Hierarchical Clustering and the Optimal Fitting
           of Ultrametrics 36
      3.2.2 Constrained Hierarchical Clustering 45

4 **Object Sequencing and Seriation** 51
   4.1 Optimal Sequencing of a Single Object Set 53
      4.1.1 Symmetric One-Mode Proximity Matrices 54
      4.1.2 Skew-Symmetric One-Mode Proximity Matrices 62
      4.1.3 Two-Mode Proximity Matrices 69
      4.1.4 Object Sequencing for Symmetric One-Mode
           Proximity Matrices Based on the Construction
           of Optimal Paths 73
   4.2 Sequencing an Object Set Subject to Precedence Constraints . . 78
   4.3 Construction of Optimal Ordered Partitions 83

5 **Heuristic Applications of the GDPP** 91
   5.1 Cluster Analysis 92
   5.2 Object Sequencing and Seriation 97

6 **Extensions and Generalizations** 103
   6.1 Introduction 103

	6.1.1	Multiple Data Sources 104
	6.1.2	Multiple Structures 107
	6.1.3	Uses for the Information in the Sets $\Omega_1,\ldots,\Omega_K$ 108
	6.1.4	A Priori Weights for Objects and/or Proximities 111
6.2	Prospects 113

Appendix: Available Programs 115

Bibliography 149

Author Index 157

Subject Index 159

Preface

The first part of this monograph's title, *Combinatorial Data Analysis* (CDA), refers to a wide class of methods for the study of relevant data sets in which the arrangement of a collection of objects is absolutely central. Characteristically, CDA is involved either with the identification of arrangements that are optimal for a specific representation of a given data set (usually operationalized with some specific loss or merit function that guides a combinatorial search defined over a domain constructed from the constraints imposed by the particular representation selected), or with the determination in a confirmatory manner of whether a specific object arrangement given *a priori* reflects the observed data. As the second part of the title, *Optimization by Dynamic Programming*, suggests, the sole focus of this monograph is on the identification of arrangements; it is then restricted further, to where the combinatorial search is carried out by a recursive optimization process based on the general principles of dynamic programming. For an introduction to confirmatory CDA without any type of optimization component, the reader is referred to the monograph by Hubert (1987). For the use of combinatorial optimization strategies other than dynamic programming for some (clustering) problems in CDA, the recent comprehensive review by Hansen and Jaumard (1997) provides a particularly good introduction.

Our purpose in writing this monograph is to provide an applied documentation source, as well as an introduction to a collection of associated computer programs that would be of interest to applied statisticians and data analysts but also accessible to a notationally sophisticated but otherwise substantively focused user. Such a person would typically be most interested in analyzing a specific data set by implementing the flexible dynamic programming method for any of a number of seemingly diverse problems encountered in CDA. The background we have tried to assume is at the same level as that required to follow the documentation for good, commercially available optimization subroutines, such as the Numerical Algorithms Group (NAG) Fortran subroutine library, or at the level of one of the standard texts in applied multivariate analysis usually used for a graduate second-year methodology course in the behavioral or social sciences. An excellent example of the latter would be the widely used text now in its fourth edition by Johnson and Wichern (1998). Draft versions of the current monograph have been used as supplementary material for a course relying on the latter text as the primary reference.

The content of the monograph itself and how the various parts are organized can be discussed under a number of headings that serve to characterize both the type of object arrangements to be identified and the form of the data on which the identification is to be based. Chapter 1 is a short preview that introduces the general topic by noting areas in combinatorial data analysis that can be approached by the optimization strategy of dynamic programming, and that presents a number of data sets to be used throughout the remaining chapters. The second chapter introduces the general dynamic programming paradigm (the GDPP, for short) and gives an introductory example of its usage in the well-known linear assignment task. The next two chapters focus the GDPP on topics within Cluster Analysis (Chapter 3) and Object Sequencing and Seriation (Chapter 4). Chapter 3 is further subdivided by several dichotomies: whether the clustering involves a single object partition (partitioning) or a hierarchy of nested partitions (and the associated representing ultrametric); the presence or absence of constraints on the type of partitions sought (typically through subsets contiguous with respect to some object order); the form of the available data with the usual distinction of having proximities between objects from a single set (one-mode) or between objects from two sets (two-mode). Chapter 4 can also be characterized by several dichotomies: whether the one-mode proximities are symmetric or skew-symmetric, with the latter representing dominance information among the objects, or whether the proximities are initially one- or two-mode. In addition, several related topics are introduced: sequencing through the construction of optimal paths (linear and circular); the incorporation of precedence constraints in the construction of an optimal order; and unifying the general areas of clustering and sequencing by identifying optimal partitions of an object set in which the classes are themselves ordered. Chapter 5 extends the GDPP heuristically for use with large(r) object sets in both the clustering and sequencing context, while (unfortunately) removing the absolute guarantee of optimality for the identified object arrangements. Finally, Chapter 6 provides preliminary discussion of a number of areas of extension and generalization that are now being pursued by the current authors and others.

An appendix is included as a user's manual for a collection of programs available as freeware on the World Wide Web (WWW) that carry out the various optimization tasks and which can be used to reproduce all the numerical examples given. We provide both the original code (in Fortran90) and executable programs (for 32-bit Intel-compatible processors running under Windows NT/95/98). Finally, we point out the liberal use throughout of chapter endnotes (rather than the more typographically intrusive footnotes). These serve several purposes: to note how some topic might be approached with one of the programs discussed in the appendix; to provide a little more peripheral comment on a topic; or to respond to a referee of an earlier version of this monograph who called for a more detailed presentation of a specific topic that we didn't include in the actual text.

PREFACE

The research reported here has been partially supported by the National Science Foundation through Grant No. SES-981407 (to Hubert) and by the Netherlands Organization for Scientific Research (NWO) through Grant No. 575-67-053 for the 'PIONEER' project 'Subject Oriented Multivariate Analysis' (to Meulman).

Lawrence Hubert
Phipps Arabie
Jacqueline Meulman

March, 2001

Chapter 1

Introduction

Many of the data analysis tasks arising naturally in the general field of classification can be given some type of combinatorial characterization that involves the identification of object groupings, partitions, or sequences. Because of this explicit usage of a combinatorial structure for data representation and interpretation, it is now common to adopt the generic term of combinatorial data analysis (CDA) to refer to the broad class of data analysis methods that variously depend on (or are directed toward the search for) such combinatorial entities (for a recent general review of CDA, see Arabie and Hubert (1996)). The present monograph falls within this area of CDA and is concerned with the identification of a variety of combinatorial structures intended to satisfy certain properties of optimality that are defined using data we have on the objects. Ultimately, the structures identified provide a mechanism to help interpret the patterns of relationship(s) reflected in the data that may exist among the objects. Also, it is usual to specify the optimality properties desired (even if only implicitly) according to whether or not a small [or large] value is achieved for some loss [or merit] criterion, where the latter characterizes numerically each possible combinatorial structure in the domain of search. Although an obvious solution strategy exists—namely, complete enumeration of all possible combinatorial structures along with their criterion values and retention of only the best—that approach will typically be computationally infeasible. For all but the smallest object sets, the search domains for the combinatorial structures of interest will be so enormous that an exhaustive search strategy is beyond the capabilities of current computational equipment.

The basic computational difficulties encountered in optimizing over combinatorial domains have spawned several well-developed areas of research. One of these areas concerns the development and evaluation of *heuristic* optimization methods that, although not guaranteed to lead to the absolute best solutions achievable, might nevertheless generate satisfactory solutions to the problem at hand (e.g., see the volume edited by Reeves (1993)). A second is involved with the characterization and classification of optimization tasks that are (inherently) difficult to solve (viz., the theory of NP-completeness; see the recent review by Day (1996)). A third is the development of (partial enumeration) methods for the exact solution of specific combinatorial optimization tasks that are more computationally efficient than complete enumeration, although there is always some limit on the absolute size of problems that can be approached (e.g., the general paradigm of branch-and-bound; see Nemhauser and Wolsey (1988)). It is within this last area of partial enumeration methods that the current monograph focuses, specifically on a very general optimization strategy called dynamic programming (DP) and the construction of recursive procedures that can allow the exact solution of certain combinatorial optimization tasks encountered in the field of classification.

The application of DP methods to data analysis problems in classification is hardly new, and a collection of precedents can be given: Fisher (1958), Held and Karp (1962), Jensen (1969), Rao (1971), Hartigan (1975, Chapter 5), Delcoigne and Hansen (1975), Adelson, Norman, and Laporte (1976), Hubert and Golledge (1981), Kruskal (1983), Hubert and Arabie (1986), Batagelj, Korenjak-Černe, and Klavžar (1994), among others. At one level our intent is to provide a comprehensive and self-contained review delineating a very general DP paradigm or schema that can in turn serve two functions. First, it may be applied in various special forms to encompass all previously proposed applications suggested in the classification literature. Second, it leads directly to several more novel uses that until now have not been noted explicitly, e.g., to certain (restricted) forms of partitioning and hierarchical clustering, or to the task of unidimensional unfolding, characterized by the joint sequencing of two object sets along a continuum. As noted above, and as mentioned and emphasized throughout this monograph, there will always be severe limits on the size of the optimization tasks that can be handled through the general DP paradigm with guaranteed optimality and with readily available computational resources. Nevertheless, we can still move far beyond the very trivial object set sizes that might be approached through complete enumeration; moreover, for some very specific contexts, it may be possible to restrict or limit the combinatorial search domain in some way (e.g., through another preliminary analysis strategy), making it feasible to consider fairly large object sets. Several examples of such reduction will be given throughout our presentation. It is also possible to apply the DP paradigm heuristically (as developed in greater detail in Chapter 5) for several optimization tasks that may exceed the usual DP strategy's capacity to guarantee absolute optimality for the specific problem at hand.

The broad outline of this monograph is as follows. Chapter 2 introduces a very general DP paradigm (Section 2.2), but does so through a simple expository example of what is called the linear assignment task (Section 2.1). Chapters 3 and 4 develop a variety of specializations of the general DP paradigm, with Chapter 3 emphasizing cluster analysis broadly defined, and Chapter 4 concentrating on the sequencing of objects along a continuum. Chapter 5 develops the use of the DP paradigm in heuristic approaches to selected problems previously identified in Chapters 3 and 4 that may be beyond exact solution by DP because of their size. The concluding Chapter 6 offers several more general observations and notes a few possible future extensions and applications.[1]

Table 1.1: *A proximity matrix from Shepard, Kilpatric, and Cunningham (1975) on the pairwise dissimilarities among the first ten single-digit integers* $\{0, 1, 2, \ldots, 9\}$ *considered as abstract concepts, averaged over raters and conditions.*

digit	0	1	2	3	4	5	6	7	8	9
0	x	.421	.584	.709	.684	.804	.788	.909	.821	.850
1		x	.284	.346	.646	.588	.758	.630	.791	.625
2			x	.354	.059	.671	.421	.796	.367	.808
3				x	.413	.429	.300	.592	.804	.263
4					x	.409	.388	.742	.246	.683
5						x	.396	.400	.671	.592
6							x	.417	.350	.296
7								x	.400	.459
8									x	.392
9										x

Because of the wide scope of the present monograph, we must leave to other, substantive sources a more extensive collection of data analysis illustrations showing the various specializations of the DP paradigm identified and discussed in the following chapters. However, to give numerical examples, we will consider a number of small data sets. Several are specific to particular optimization contexts and are thus introduced during these discussions; here, we present three small object sets used more generally throughout the monograph. The first, given as a dissimilarity matrix in Table 1.1, is taken from Shepard, Kilpatric, and Cunningham (1975). The stimulus domain is the first ten single-digits $\{0, 1, 2, \ldots, 9\}$ considered as abstract concepts, and the proximity matrix was constructed by averaging dissimilarity ratings for distinct pairs of those integers over a number of subjects and conditions (thus, given the dissimilarity interpretation for these proximity values, smaller entries in the table reflect more similar digits). A direct inspection of these data suggests there may be some very regular but possibly complex manifest patterning reflecting either structural characteristics of the digits (e.g., the powers of 2 or of 3, the salience of the two additive/multiplicative identities [0/1], oddness/evenness) or of absolute magnitudes. These data will be relied on in Chapter 3 to provide concrete numerical illustrations of various aspects of a clustering task.

A second data set used in Chapter 4 to illustrate several optimization problems involved with sequencing an object set along a continuum is a very old one, originally collected in 1929 in a study of the influence of motion pictures on children's attitudes (see Thurstone (1959), pp. 309–319). Both before and after seeing a film entitled *Street of Chance*, which depicted the life of a gambler, 240 school children were asked to compare the relative seriousness of 13 offenses presented in all 78 possible pairs: bankrobber, gambler, pickpocket, drunkard, quack doctor, bootlegger, beggar, gangster, tramp, speeder, petty thief, kidnapper, and smuggler. The data are given in Table 1.2, where the

Table 1.2: *The proportions of school children who evaluate the column offense as more serious than the row offense (taken from Thurstone (1959), p. 311). The above-diagonal entries are before showing the film* Street of Chance; *those below the diagonal were collected after viewing the motion picture. Apparently, the single pair (petty thief, kidnapper) was inadvertently not presented for evaluation, although no mention of this anomaly is made in Thurstone (1959). The values of .98 for (a) and .03 for (b) are imputed based on an assumption of strong stochastic transitivity (e.g., see Bezembinder and van Acker (1980)) and the observed proportions for the two pairs (petty thief, bankrobber) and (kidnapper, bankrobber).*

offense	1	2	3	4	5	6	7	8	9	10	11	12	13
1:bankrobber	x	.07	.08	.05	.27	.29	.01	.50	.00	.06	.02	.73	.21
2:gambler	.79	x	.71	.52	.76	.92	.07	.92	.05	.41	.49	.90	.81
3:pickpocket	.93	.51	x	.25	.67	.75	.02	.86	.02	.39	.42	.87	.68
4:drunkard	.95	.70	.70	x	.81	.95	.01	.92	.03	.37	.62	.91	.87
5:quack doctor	.67	.36	.28	.16	x	.49	.02	.70	.02	.12	.22	.64	.55
6:bootlegger	.70	.31	.30	.13	.50	x	.00	.79	.01	.09	.26	.68	.50
7:beggar	.98	.95	.97	.94	.98	.98	x	.96	.42	.86	.96	1.0	.99
8:gangster	.50	.18	.13	.11	.32	.27	.01	x	.02	.08	.08	.36	.31
9:tramp	1.0	.96	.98	.96	.99	.98	.64	.99	x	.91	.97	.99	1.0
10:speeder	.94	.73	.68	.67	.89	.90	.21	.94	.13	x	.58	.90	.92
11:petty thief	.97	.64	.62	.47	.81	.76	.06	.89	.05	.36	x	(a)	.78
12:kidnapper	.38	.27	.16	.08	.35	.30	.02	.62	.01	.08	(b)	x	.27
13:smuggler	.73	.31	.30	.16	.46	.49	.02	.66	.02	.11	.24	.64	x

entries show the proportion of the school children who rated the offense listed in the column to be more serious than the offense listed in the row. The above-diagonal entries were obtained before the showing of the film; those below were collected after. The obvious substantive question here involves the effect of the film on the assessment of the offense of being a gambler.

A third data matrix given in Table 1.3 was originally collected by Marks (1965) and has been reanalyzed elsewhere (e.g., Schiffman, Reynolds, and Young (1981); Schiffman and Falkenberg (1968); Hubert and Arabie (1995a)). These data refer to the absorption of light in a goldfish retina at specific wavelengths and by various cones, and are provided here as dissimilarities in Table 1.3, defined by 200 minus the measured heights of these ordinates for the various spectral frequencies (see Schiffman, Reynolds, and Young (1981), p. 329). The substantive issue concerns the sensitivity of certain receptors (cones) to specific wavelengths.

Endnote

[1]There are a number of computer programs that are used in the remaining chapters to carry out the various special cases of the DP paradigm reviewed

Table 1.3: *Dissimilarities between specific receptors and wavelengths in goldfish retina (taken from Schiffman, Reynolds, and Young (1981), p. 329).*

wave-length: receptor	1 green (530)	2 yellow (585)	3 red (660)	4 b-indigo (458)	5 b-green (498)	6 blue (485)	7 green (540)	8 orange (610)	9 violet (430)
1	103	63	155	198	148	154	94	108	186
2	46	107	200	99	60	78	47	156	101
3	188	200	200	47	143	111	196	200	53
4	48	84	174	115	73	97	52	125	154
5	114	61	54	141	148	142	121	47	113
6	49	91	200	122	79	115	46	143	127
7	116	49	80	135	127	123	98	46	156
8	186	200	200	48	100	75	200	200	55
9	168	177	200	46	125	90	176	183	47
10	145	80	68	200	161	160	138	53	200
11	144	64	89	173	176	177	128	56	140

and that are denoted by various names in the course of our presentation (e.g., DPCL1U, DPCL1R, DPHI1U, among others). Executable versions (for Intel processors) of all the programs mentioned can be obtained from a World Wide Web site set up by the authors, as well as the original source code for the programs written in Fortran90. A computational appendix to this monograph (ftp://www.psych.uiuc.edu/pub/hubert) provides a summary listing of all the available programs (along with a review of what the various acronyms refer to). The body of the appendix gives a discussion of their use, as well as illustrative input and output. In each instance, there are upper limits on the absolute size of the problems that can be approached with a particular program because of realistic upper-bounds on the availability of the necessary random access memory (RAM) required for storage (and these approximate limits will be noted in our discussion). However, because none of the programs rely on large fixed-size arrays that must be declared initially, but instead use Fortran90's ability to allocate arrays dynamically, the program itself will decide (and so inform the user) whether sufficient RAM exists in the system on which the program is being run to solve a particular problem.

Chapter 2

General Dynamic Programming Paradigm

We wish to introduce the General Dynamic Programming Paradigm (GDPP) in a notational form sufficiently flexible for the variety of optimization tasks surveyed in the remaining chapters of this monograph. Thus, it is convenient to have a preliminary example of how a simply stated (and also well-known) optimization task can be solved by DP. The elements introduced for its solution can then be used to give concrete referents for the more general notational system of the GDPP explicated in Section 2.2. We thus begin by giving a DP solution to the linear assignment problem.

2.1 An Introductory Example: Linear Assignment

The linear assignment (LA) task (in one of its simple variants) can be phrased using two object sets, where each contains n members, say, $U = \{u_1, \ldots, u_n\}$ and $V = \{v_1, \ldots, v_n\}$, and an $n \times n$ merit matrix $\mathbf{C} = \{c_{ij}\}$, where c_{ij} denotes the value of pairing objects u_i and v_j. The optimization task is to find a one-to-one matching of the objects in U and V that will maximize the sum of the n merit values produced by the matching. Interpretively, the objects in U might represent n people who must be allocated to the n tasks denoted by the objects in V, where c_{ij} is the value of assigning task v_j to person u_i.[2]

A complete enumeration strategy for the LA task requires the evaluation of the sum of merit values for all $n!$ one-to-one matchings of the objects in U and V. Explicitly, if $\rho(\cdot)$ denotes a permutation of the first n integers representing the n integer subscripts on the objects in U (i.e., $\rho(i) = j$ if v_j is assigned to u_i), and therefore defining a one-to-one matching between U and V, the sum of the merit values for that matching can be represented as

$$\Gamma(\rho(\cdot)) = \sum_{i=1}^{n} c_{i\rho(i)}.$$

Thus, $\Gamma(\rho(\cdot))$ could be evaluated for each of the $n!$ permutations, and an optimal solution to the LA problem identified by the largest value achieved for the index.

To approach the LA task through a recursive strategy, avoiding the need to enumerate completely all $n!$ one-to-one matchings of U to V, suppose a collection of n sets is first defined as $\Omega_1, \ldots, \Omega_n$, where Ω_k (for $1 \le k \le n$) contains all subsets that have exactly k of the n integer subscripts indexing the objects in V. If A_k denotes a subset contained in Ω_k, and therefore a subset of size k of the first n integers, let $\mathcal{F}(A_k)$ be the largest sum of k merit values that could be achieved by assigning the k indices present in A_k, in some order, to the first k objects $u_1, \ldots, u_k$. For $A_1 \in \Omega_1$, and where A_1 includes a single subscript, say, h (i.e., $A_1 = \{h\}$), $\mathcal{F}(A_1) = c_{1h}$. Consequently, $\mathcal{F}(A_1)$ may be obtained directly for all $A_1 \in \Omega_1$, and starting with these initial evaluations, the optimal sums of merit values $\mathcal{F}(A_k)$ for $k = 2, \ldots, n$ can be constructed recursively as

$$\mathcal{F}(A_k) = \max_{h \in A_k}[\mathcal{F}(A_k - \{h\}) + c_{kh}], \quad (2.1)$$

where $A_k \in \Omega_k$ and $A_k - \{h\} \in \Omega_{k-1}$ when $h \in A_k$. Starting from $\mathcal{F}(A_1)$ for all $A_1 \in \Omega_1$, we first obtain $\mathcal{F}(A_2)$ for all $A_2 \in \Omega_2$; then, based on $\mathcal{F}(A_2)$ for all $A_2 \in \Omega_2$, we obtain $\mathcal{F}(A_3)$ for all $A_3 \in \Omega_3$, and so on, until we finally reach $\mathcal{F}(A_n)$. Because there is only a single subset A_n in Ω_n and this subset contains all n integers, the optimal value for the LA task must be $\mathcal{F}(\{1, 2, \ldots, n\})$. An actual assignment that would produce the optimal value can be obtained by working backwards through the recursive process; i.e., the last assigned subscript (which is matched to object u_n) that led to $\mathcal{F}(\{1, 2, \ldots, n\})$ is identified and removed from the set, leaving a subset in Ω_{n-1}, say, A'_{n-1}; the last assigned subscript (which is matched to object u_{n-1}) that produced $\mathcal{F}(A'_{n-1})$ is identified and removed from A'_{n-1}, producing a subset in Ω_{n-2}, and so on, until a complete assignment is identified by reaching a subset in Ω_1 that includes a single subscript.

The explicit justification for the appropriateness of a recursive system, such as that in (2.1), is usually made in the following form (e.g., see Held and Karp (1962), p. 197): the value $\mathcal{F}(A_k)$ denotes the optimal sum of merit indices achievable by assigning the k subscripts in A_k to the first k objects $u_1, \ldots, u_k$, where one of these subscripts, call it h, must be assigned to u_k and the remaining subscripts in $A_k - \{h\}$ *must be assigned optimally* to $u_1, \ldots, u_{k-1}$. The total sum of merit indices for such an assignment is $\mathcal{F}(A_k - \{h\}) + c_{kh}$, and taking the maximum over all possible choices for h leads to the recursion in (2.1). The obvious key to this argument is the italicized condition, which is satisfied because the additive increment of c_{kh} generated by assigning the subscript h to u_k does *not* depend on the order in which the subscripts in $A_k - \{h\}$ are assigned to $u_1, \ldots, u_{k-1}$. (An illustration will be given in Section 4.1.3 on

2.1. INTRODUCTORY EXAMPLE: LINEAR ASSIGNMENT

unidimensional unfolding where, for a particularly natural measure that we might seek to optimize, a violation of such a condition follows; therefore, no obvious DP approach is possible using that specific measure.)

A numerical example. To give a small numerical illustration for clarification, suppose U and V each contain four objects and the merit indices are given by the 4×4 matrix $\mathbf{C}$:

	v_1	v_2	v_3	v_4
u_1	5	4	2	10
u_2	14	9	13	7
u_3	3	8	1	12
u_4	15	6	11	16

A schematic of the recursive process from (2.1) appears in Figure 2.1, which traces the construction of an optimal assignment of the four subscripts on the objects in V to the four subscripts on the objects in U considered (without loss of generality) in the order $1 \to 2 \to 3 \to 4$. The enumeration of all the $2^4 - 1 = 15$ nonempty subscript subsets within the sets $\Omega_1, \ldots, \Omega_4$ is provided along with the optimal values $\mathcal{F}(A_1), \ldots, \mathcal{F}(A_4)$ beside each subset. These optimal values, as noted in conjunction with the recursion in (2.1), are obtained as follows: first the values $\mathcal{F}(A_1)$ for subsets A_1 in Ω_1 containing single objects are found by simple inspection of the entries in the first row of the matrix $\mathbf{C}$ that gives the merit values for assigning one of the four objects in V to u_1:

$$\mathcal{F}(\{1\}) = 5, \ \mathcal{F}(\{2\}) = 4, \ \mathcal{F}(\{3\}) = 2, \ \mathcal{F}(\{4\}) = 10.$$

Based on these values of $\mathcal{F}(A_1)$ for the single-object subsets $A_1 \in \Omega_1$, the values for $\mathcal{F}(A_2)$ for the dyadic subsets $A_2 \in \Omega_2$ are obtained by the recursion in (2.1):

$$\mathcal{F}(\{1,2\}) = \max(\mathcal{F}(\{1,2\} - \{1\}) + c_{21}, \mathcal{F}(\{1,2\} - \{2\}) + c_{22}) = \max(4+14, 5+9) = 18;$$

$$\mathcal{F}(\{1,3\}) = \max(\mathcal{F}(\{1,3\} - \{1\}) + c_{21}, \mathcal{F}(\{1,3\} - \{3\}) + c_{23}) = \max(2+14, 5+13) = 18;$$

$$\mathcal{F}(\{1,4\}) = \max(\mathcal{F}(\{1,4\} - \{1\}) + c_{21}, \mathcal{F}(\{1,4\} - \{4\}) + c_{24}) = \max(10+14, 5+7) = 24;$$

$$\mathcal{F}(\{2,3\}) = \max(\mathcal{F}(\{2,3\} - \{2\}) + c_{22}, \mathcal{F}(\{2,3\} - \{3\}) + c_{23}) = \max(2+9, 4+13) = 17;$$

$$\mathcal{F}(\{2,4\}) = \max(\mathcal{F}(\{2,4\} - \{2\}) + c_{22}, \mathcal{F}(\{2,4\} - \{4\}) + c_{24}) = \max(10+9, 4+7) = 19;$$

$$\mathcal{F}(\{3,4\}) = \max(\mathcal{F}(\{3,4\} - \{3\}) + c_{23}, \mathcal{F}(\{3,4\} - \{4\}) + c_{24}) = \max(10+13, 2+7) = 23.$$

$\Omega_1 : \mathcal{F}(A_1)$ $\qquad\qquad \Omega_2 : \mathcal{F}(A_2)$ $\qquad\qquad \Omega_3 : \mathcal{F}(A_3)$ $\qquad\qquad \Omega_4 : \mathcal{F}(A_4)$

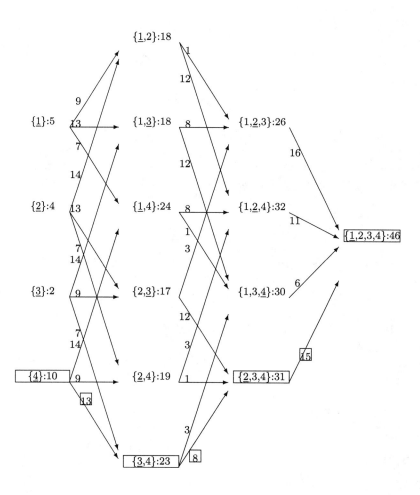

Figure 2.1: *A graphical representation for the recursive solution of the illustrative 4×4 linear assignment task presented in the text. The numbers immediately following colons are either entries or sums of entries in the data matrix* **C**. *Underlined numbers refer to the most recent addition to a subset.*

Continuing, and using these latter results for $\mathcal{F}(A_2)$ when $A_2 \in \Omega_2$, the evaluations of $\mathcal{F}(A_3)$ for the three-object subsets $A_3 \in \Omega_3$ are

$$\mathcal{F}(\{1,2,3\}) = \max(\mathcal{F}(\{1,2,3\} - \{1\}) + c_{31}, \mathcal{F}(\{1,2,3\} - \{2\}) + c_{32},$$
$$\mathcal{F}(\{1,2,3\} - \{3\}) + c_{33}) = \max(17+3, 18+8, 18+1) = 26;$$

2.1. INTRODUCTORY EXAMPLE: LINEAR ASSIGNMENT

$$\mathcal{F}(\{1,2,4\}) = \max(\mathcal{F}(\{1,2,4\} - \{1\}) + c_{31}, \mathcal{F}(\{1,2,4\} - \{2\}) + c_{32},$$
$$\mathcal{F}(\{1,2,4\} - \{4\}) + c_{34}) = \max(19 + 3, 24 + 8, 18 + 12) = 32;$$

$$\mathcal{F}(\{1,3,4\}) = \max(\mathcal{F}(\{1,3,4\} - \{1\}) + c_{31}, \mathcal{F}(\{1,3,4\} - \{3\}) + c_{33},$$
$$\mathcal{F}(\{1,3,4\} - \{4\}) + c_{34}) = \max(23 + 3, 24 + 1, 18 + 12) = 30;$$

$$\mathcal{F}(\{2,3,4\}) = \max(\mathcal{F}(\{2,3,4\} - \{2\}) + c_{32}, \mathcal{F}(\{2,3,4\} - \{3\}) + c_{33},$$
$$\mathcal{F}(\{2,3,4\} - \{4\}) + c_{34}) = \max(23 + 8, 19 + 1, 17 + 12) = 31.$$

Finally, $\mathcal{F}(A_4)$ for the single four-object subset $A_4 = \{1,2,3,4\} \in \Omega_4$ is obtained from $\mathcal{F}(A_3)$ when $A_3 \in \Omega$:

$$\mathcal{F}(\{1,2,3,4\}) = \max(\mathcal{F}(\{1,2,3,4\} - \{1\}) + c_{41}, \mathcal{F}(\{1,2,3,4\} - \{2\}) + c_{42},$$
$$\mathcal{F}(\{1,2,3,4\} - \{3\}) + c_{43}, \mathcal{F}(\{1,2,3,4\} - \{4\} + c_{44}) =$$
$$\max(31 + 15, 30 + 6, 32 + 11, 26 + 16) = 46.$$

The increments in merit are listed on the directed lines that show a transition from a subset $A_k \in \Omega_k$ to a subset $A_{k+1} \in \Omega_{k+1}$, $1 \le k \le 3$. For each subset in Ω_k an indication is given by an underline as to which subscript was the last assigned in generating the optimal value for that subset. Based on the information in Figure 2.1 and working backwards from the optimal value of 46 for $\mathcal{F}(\{\underline{1},2,3,4\})$, where $\underline{1}$ was the last assigned subscript, we successively reach the subsets $\{\underline{2},3,4\}$, $\{\underline{3},4\}$, and $\{\underline{4}\}$. Thus, the optimal matching is defined by the permutation $\rho(1) = 4$, $\rho(2) = 3$, $\rho(3) = 2$, $\rho(4) = 1$, i.e., the one-to-one matching is $u_1 \leftrightarrow v_4$, $u_2 \leftrightarrow v_3$, $u_3 \leftrightarrow v_2$, $u_4 \leftrightarrow v_1$. The specific path followed in the figure defining the optimal matching is marked by placing boxes around the elements in $\Omega_1, \ldots, \Omega_4$ that are successively visited; boxes are also placed around the transition increments used to obtain the optional merit sum of 46.

The simple example just described can be used to illustrate several general characteristics of DP methods and introduce some common terminology that will appear throughout what follows.

(1) When it is possible to carry out a DP approach for a particular combinatorial optimization task, the amount of computational effort is usually reduced substantially, compared to that for a complete enumeration. In the LA task, for example, complete enumeration requires an evaluation of the index $\Gamma(\cdot)$ over all $n!$ possible permutations of the subscripts $1, 2, \ldots, n$. Through (2.1), only the values $\mathcal{F}(\cdot)$ need be obtained for $2^n - 1$ nonempty subsets of the set of subscripts.

(2) Although $2^n - 1$ is smaller than $n!$ for $n \ge 4$ and $n!$ grows at a much faster rate than $2^n - 1$, the latter number in an absolute sense is quite large as n becomes large. This growth does severely limit the size of the problems that can be effectively approached with a DP strategy when used to obtain an optimal solution. Required for setting up a DP method for the LA task is a means to save and access *all* the values of $\mathcal{F}(\cdot)$ that are recursively generated over all nonempty subsets of a set containing n objects, in addition to a storage mechanism for keeping track of the last subscripts that led to these optimal

values. Thus, large array capacity must be available for implementing all the various DP methods reviewed as special instances of the general paradigm.[3]

We might also note at this point that throughout the current monograph we will generally be content with only obtaining a single optimal solution for a particular optimization task, and will not deal explicitly with the issue of identifying *all* possible optimal solutions. To do so in the LA task, for instance, would require a storage mechanism for keeping track of every last assigned object that would lead to the same optimal values, $\mathcal{F}(A_k)$, at each stage of the recursion, and eventually identifying all optimal solutions by working backwards through the recursion and considering all the alternative paths back to Ω_1. Although such a process would be possible to carry out, it would also require the dedication of much more (and usually scarce) storage capacity. Because of this increase, the possible nonuniqueness of an optimal solution will typically be left unaddressed, both here and for uses of the GDPP introduced in the next section.

(3) The LA task was phrased as a maximization problem, but it could be restated as the minimization of cost (or loss), where an optimal assignment would now have minimum cost, i.e., the sum of the n cost values induced by an assignment is to be minimized. This observation will be true for all specializations of the GDPP presented in Chapters 3 and 4, and depending on the problem's definition, either a minimum or a maximum value will be sought. As a more basic alteration to the recursion in (2.1), the additive combination of c_{kh} and $\mathcal{F}(A_k - \{h\})$ could be replaced by the minimum of c_{kh} and $\mathcal{F}(A_k - \{h\})$ to develop a recursion of the form

$$\mathcal{F}(A_k) = \max_{h \in A_k}[\min(\mathcal{F}(A_k - \{h\}), c_{kh})] \qquad (2.2)$$

for $2 \leq k \leq n$. Thus, an optimal assignment is now one that maximizes the minimum merit value over all n such merit values induced by the assignment. Alternatively, we might minimize the maximum cost value if the entries in the matrix $\{c_{ij}\}$ represent cost rather than merit. In general, all the various DP specializations in what follows, analogous to (2.2), have a max/min (or min/max) form.

(4) Although the LA task may serve as a good introductory illustration for how a DP recursion can be developed, it should be noted that at least for this specific combinatorial optimization task, alternative and generally much better methods exist for an optimal solution according to both the size of the problems that can be handled and the computational effort required. For example, LA can be phrased as a linear programming problem, and thus all the computational routines for linear programming could be applied directly, such as the well-known simplex algorithm. Furthermore, given the very special structure of LA when reformulated as a linear programming problem, more efficient methods than, say, the general simplex algorithm have been developed; e.g., see the discussion of the Hungarian procedure in Hillier and Lieberman (1990, pp. 240–243). Generally, for the various specializations of the general DP strategy we

will discuss, such alternative and more efficient methods are not available, in contrast with the LA task.

(5) There is a very convenient graph-theoretic interpretation for the recursive processes given in (2.1) and (2.2) through the construction of 'longest' paths in an acyclic directed graph.[4] Such an interpretation may be very helpful, at least intuitively, for a better understanding of the basic problem the recursive processes solve (similar graphical representations could be given for all the applications of the GDPP reviewed in the chapters to follow). For example, suppose that in Figure 2.1 a (source) node is first placed at the far left and directed lines all having a weight of zero are drawn to each of the four nodes corresponding to the entities in Ω_1. If the 'length' of a path from this source node to the (sink) node defined by the set $S = \{1, 2, 3, 4\}$ is operationalized as either the sum of the weights for the directed lines along the path or, alternatively, as the minimum weight for the directed lines along the path, the recursions in (2.1) and (2.2) identify a longest path from the source node to the sink node. In our case, for instance, using the sum of weights as our characterization of 'length,' the longest path is 46, which is the same as the value attached to $\mathcal{F}(\{1, 2, 3, 4\})$. (Because transition costs from one node to the next obviously do not depend on how the first node was reached, the short argument for the validity of the recursions given earlier may be intuitively more compelling within this general type of graphical framework.) Analogously, if a cost interpretation is given for transitions between nodes, and if either the sum of the weights or the maximum weight along a path is now to be minimized, the alternatives to (2.1) and (2.2) using minimization would solve a shortest path problem from the source node to the sink node.

2.2 The GDPP

To present the GDPP corresponding in one special case to the specific LA recursion given in (2.1) and based on maximization and the use of additive increments in merit, we first define a collection of K sets of entities[5] $\Omega_1, \ldots, \Omega_K$, where it is possible by some operation to transform entities in Ω_{k-1} to certain entities in Ω_k for $2 \leq k \leq K$. Each such transformation can be assigned a merit value based *only* on the entity in Ω_{k-1} and the transformed entity in Ω_k. An entity in Ω_k is denoted by A_k, and $\mathcal{F}(A_k)$ is the optimal value that can be assigned to A_k based on the sum of the merit increments necessary to transform an entity in Ω_1, step-by-step, to $A_k \in \Omega_k$. If $A_{k-1} \in \Omega_{k-1}$ can be transformed into $A_k \in \Omega_k$, the merit of that single transition will be denoted by $M(A_{k-1}, A_k)$, where the latter *does not depend* on how A_{k-1} may have been reached starting from an entity in Ω_1. Given these conditions, and assuming the values $\mathcal{F}(A_1)$ for $A_1 \in \Omega_1$ are available to initialize the recursive system, $\mathcal{F}(A_k)$ may be constructed for $k = 2, \ldots, K$ as

$$\mathcal{F}(A_k) = \max[\mathcal{F}(A_{k-1}) + M(A_{k-1}, A_k)], \quad (2.3)$$

where $A_k \in \Omega_k$, $A_{k-1} \in \Omega_{k-1}$, and the maximum is taken over all A_{k-1} that can be transformed into A_k.[6]

The specialization of (2.3) to the LA recursion in (2.1) is very direct, merely by interpreting K to be n and letting the entities in Ω_k be all subsets of the subscript set $\{1, 2, \ldots, n\}$ that contain k members. A transformation is possible between A_{k-1} and A_k if $A_{k-1} \subset A_k$, i.e., A_{k-1} and A_k differ by one subscript; the increment, $M(A_{k-1}, A_k)$, is the merit of assigning the single subscript in $A_k - A_{k-1}$ to the kth entry u_k in U. Also, as mentioned in the context of the LA task, a minimization analogue of (2.3) would be immediate using a cost of transforming A_{k-1} to A_k, $C(A_{k-1}, A_k)$:

$$\mathcal{F}(A_k) = \min[\mathcal{F}(A_{k-1}) + C(A_{k-1}, A_k)], \tag{2.4}$$

or to max/min and min/max forms:

$$\mathcal{F}(A_k) = \max[\min(\mathcal{F}(A_{k-1}), M(A_{k-1}, A_k))], \tag{2.5}$$

$$\mathcal{F}(A_k) = \min[\max(\mathcal{F}(A_{k-1}), C(A_{k-1}, A_k))]. \tag{2.6}$$

As in (2.3), for each expression in (2.4), (2.5), and (2.6), the leading maximization or minimization is over all $A_{k-1} \in \Omega_{k-1}$ that can be transformed into A_k.

Endnotes

[2]For this latter interpretation of optimally assigning jobs or tasks to individuals, LA has a long history in the psychometric literature under the title of the personnel classification problem. The interested reader might consult Thorndike (1950) for historical context for this usage, including an anecdote, purportedly due to a mathematician, which states that because there were a finite number of possible solutions to the LA task and therefore it could be solved by complete enumeration, the problem was of no mathematical interest.

[3]In managing the storage of subset information as in the LA task, for example, we can deal effectively with n's in their lower 20's with the usual type of Pentium-level PC now available and with the amount of RAM typically contained. Throughout this monograph we have in mind that a 'commonly available storage configuration' (as of 1999, say) would be a system that contained somewhere between 64 and 128 MB of RAM.

[4]A directed graph can be characterized by a finite set of nodes (or vertices) and a set of ordered pairs of nodes (called arcs, directed edges, or directed lines). Typically, the nodes are given as points, and for each ordered pair of nodes in the second set an arrow is drawn from the first node to the second. A source node is one that only has directed lines drawn from it; a sink node is one that only has directed lines drawn to it. A path in a directed graph is an ordered sequence of nodes with associated directed lines between all adjacent pairs of nodes, where a consistency of directionality is maintained in moving from the first node to the last. An acyclic directed graph is one in which no paths exist

2.2. THE GDPP

that would include the same node more than once, i.e., no paths that are cycles are present in an acyclic directed graph.

[5] The neutral term 'entity' will be used consistently to refer to a member in one of the sets $\Omega_1, \ldots, \Omega_K$. Depending on the particular application, an entity may be any of a variety of combinatorial objects, e.g., a subset (possibly of a fixed size), a partition, an ordered pair of indices, an ordered pair consisting of a subset and a single index, and so on.

[6] In addition to providing graphical interpretations of the construction of optimal paths in directed graphs, all the applications of the GDPP we review could alternatively be rephrased using partially ordered sets and the construction of a corresponding Hasse diagram (the latter, in effect, would be the directed graph just alluded to). A single set $\Omega \equiv \bigcup_{k=1}^{K} \Omega_k$ would be constructed and a relation $\preceq$ defined on Ω to enable deciding for two members of Ω, say, A and A', whether $A \preceq A'$. (In the linear assignment task, for example, the set Ω would be all nonempty subsets of the set of subscripts $\{1, 2, \ldots, n\}$, and "$\preceq$" would be simple subset inclusion "$\subseteq$". In other applications of the GDPP, the relation "$\preceq$" would typically require a little more notational care to define precisely.) The relation $\preceq$ is a *partial order* on Ω (and the pair $(\Omega, \preceq)$ is referred to as a partially ordered set or poset) if $\preceq$ satisfies three conditions: reflexive ($A \preceq A$ for $A \in \Omega$); antisymmetric ($A \preceq A'$ and $A' \preceq A$ imply $A = A'$); and transitive ($A \preceq A'$ and $A' \preceq A''$ imply $A \preceq A''$). Given $\preceq$, define the relation $\prec$ by $A \prec A'$ if and only if $A \preceq A'$ and $A \neq A'$. The element A' is said to cover A if $A \prec A'$ and there is no A'' such that $A \prec A'' \prec A'$. A Hasse diagram of the poset $(\Omega, \preceq)$ is a figure consisting of the members of Ω with a directed line segment from A to A' whenever A' covers A (i.e., when a transition is possible from an entry in Ω_{k-1} to one in Ω_k in our original representation by an acyclic directed graph).

Chapter 3

Cluster Analysis

The two major sections of this chapter discuss, respectively, the tasks of (a) partitioning some set of n objects, $S = \{O_1, \ldots, O_n\}$, into M mutually exclusive and exhaustive (as well as nonempty) subsets, and (b) hierarchical clustering in which a sequence of hierarchically related partitions of S must be constructed. In both cases, it is assumed that some $n \times n$ symmetric proximity matrix $\mathbf{P} = \{p_{ij}\}$ is available, where p_{ij} denotes the nonnegative dissimilarity of O_i and O_j (i.e., larger values for p_{ij} indicate more dissimilar objects), and where $p_{ii} = 0$ for $1 \leq i \leq n$. Also common to both sections are possible extensions to the situation where S itself may be the union of two disjoint subsets and the only nonmissing proximities in $\mathbf{P}$ are those defined between these distinct sets,[7] and to the possibility of imposing certain admissibility criteria for the type of partitionings and/or hierarchical clusterings to be generated. In general, the use of admissibility criteria may be implemented by the judicious definition of large positive or large negative cost or merit increments, respectively, that would disallow the consideration of certain transitions between some of the entities $A_{k-1} \in \Omega_{k-1}$ and some entities $A_k \in \Omega_k$. These criteria might be defined using the proximity matrix $\mathbf{P}$ and the relationship that A_{k-1} and A_k bear to the information given in $\mathbf{P}$, or by some prior restriction to certain subsets of the originally defined sets $\Omega_1, \ldots, \Omega_K$. In this latter case, it may even be possible to redefine the recursive process using a different collection $\Omega_1, \ldots, \Omega_K$, increasing the size of the problems that might be effectively approached. Several of these admissibility issues will be considered in the following sections.

3.1 Partitioning

The most direct characterization of the partitioning task can be stated as follows: given $S = \{O_1, \ldots, O_n\}$ and $\mathbf{P} = \{p_{ij}\}$, find a collection of M mutually exclusive and exhaustive nonempty subsets of S, say, $S_1, \ldots, S_M$, such that for some measure of heterogeneity $H(\cdot)$ that attaches a value to each possible subset of

S, either the sum

$$\sum_{m=1}^{M} H(S_m)$$

or, alternatively,

$$\max[H(S_1), \ldots, H(S_M)]$$

is minimized. This stipulation assumes that heterogeneity has a cost interpretation and that smaller values of the heterogeneity indices represent the 'better' subsets (or clusters). In any event, if $H(S_m)$ for some $S_m \subseteq S$ depends only on those proximities from **P** that are within S_m and/or between S_m and $S - S_m$, an application of the GDPP given in (2.4) and (2.6) is possible. We define K to be M, and let each of the sets $\Omega_1, \ldots, \Omega_K$ contain *all* of the $2^n - 1$ nonempty subsets of the n object subscripts; $\mathcal{F}(A_k)$ is the optimal value for a partitioning into k classes of the object subscripts present in A_k. A transformation of an entity in Ω_{k-1} (say, A_{k-1}) to one in Ω_k (say, A_k) is possible if $A_{k-1} \subset A_k$, with cost $C(A_{k-1}, A_k) \equiv H(A_k - A_{k-1})$. Thus, beginning with the heterogeneity indices $H(A_1)$ for *every* subset $A_1 \subseteq S$, the recursion can be carried out, with the optimal solution represented by $\mathcal{F}(A_K)$ when $A_K = S$. The $M(\equiv K)$ subsets of S constituting the solution are obtained (as usual) by working backwards through the recursive steps. It might also be noted that besides yielding an optimal solution containing M subsets, optimal solutions for partitions having from $M - 1$ to 2 subsets can be obtained immediately from $\mathcal{F}(A_{M-1}), \ldots, \mathcal{F}(A_2)$, when $A_{M-1} = \cdots = A_2 = S$. The optimal values are generated as part of the recursion, and again, the actual optimal partitions can be identified by a process of working backwards.[8]

There is a wide variety of possible subset heterogeneity measures that could be used to obtain an optimal partitioning of S, and we are really limited only by our own ingenuity in creating them. Some are obvious, such as the ones mentioned below, but there are, for example, many other alternatives derivable from graph-theoretic concepts that all follow a general format: for some given S_m, the heterogeneity of S_m could be defined by the minimum proximity needed to ensure that a certain property exists in the subgraph defined by the objects in S_m with edges between objects present whenever their proximity is less than or equal to this minimum value. Two graph-theoretic options that fit this type of format are mentioned below. These use the notation of $\max(S_m)$ and $\min(S_m)$, which refer to subgraph completeness and connectivity, respectively.[9] For a comprehensive review of many other possibilities, see Hubert (1974a).

For notational convenience here as well as later, if $S_m \subseteq S$ contains n_m objects, $\text{sum}(S_m)$ will denote the sum of all $n_m(n_m - 1)$ proximities between objects within S_m, and $\text{sum}(S_m, S - S_m)$ the sum of all $2n_m(n - n_m)$ proximities between objects in S_m and $S - S_m$, where each distinct object pair is included twice depending on the order of the two constituent objects.[10] Also, $\max(S_m)$ will indicate the maximum proximity within S_m; $\min(S_m)$ will denote the minimum proximity value needed to ensure connectivity within S_m, i.e., a path exists between any two objects within S_m, where all proximities between adjacent object pairs along the path are not larger than the identified minimum proximity; and $\mathrm{span}(S_m)$ will be the length of the minimum spanning tree defined between the objects within S_m. The notation $\mathrm{count}\,a(S_m, S - S_m)$ will specify the number of instances in which a proximity within S_m is strictly greater than one between S_m and $S - S_m$, and $\mathrm{count}\,b(S_m, S - S_m)$ the number of times an object within S_m has a larger proximity to another object within S_m than it has to an object in $S - S_m$.[11] Given these definitions, the following twelve possible heterogeneity measures for a subset S_m are immediate. They are indexed by lower-case Roman numerals for convenience of reference both here and throughout the monograph:[12]

(i) $\mathrm{sum}(S_m)$; each subset will contribute directly according to the number of object pairs it contains, and thus, smaller size subsets will tend to be given smaller heterogeneity values.

(ii) $[1/(n_m(n_m - 1))]\mathrm{sum}(S_m)$; because the average proximity within S_m is used, the absolute subset size will not affect the assigned heterogeneity.

(iii) $[1/(2n_m)]\mathrm{sum}(S_m)$; the number of objects (in contrast to the number of object pairs as in (i)) contributes directly to heterogeneity. If the original proximities are squared Euclidean distances between numerically given vectors (or profiles) for the n objects over some set of variables, then this measure is equivalent to the sum of squared Euclidean distances between each profile and the mean profile for the subset (this is usually called the sum of squared error or the k-means criterion, e.g., see Späth (1980), p. 52).

(iv) $\max(S_m)$; this is commonly called the diameter of the cluster S_m.

(v) $\min(S_m)$; as noted earlier, this value is the minimum proximity needed to ensure connectivity in the subgraph defined by the objects in S_m. Using this measure, if the maximum heterogeneity is minimized over subsets, the same partition is identified as is obtained using the well-known single-link hierarchical clustering method and choosing that level of the hierarchy with M subsets.

(vi) $[1/(n_m(n_m - 1))]\mathrm{sum}(S_m) - [1/(2n_m(n - n_m))]\mathrm{sum}(S_m, S - S_m)$; because of the keying of proximities as dissimilarities, the average within-subset proximity minus the average between that subset and its complement should be negative and large in absolute value for highly homogeneous subsets.

(vii) $-[1/(2n_m(n - n_m))]\mathrm{sum}(S_m, S - S_m)$; in contrast to (vi), only the negative of the average proximity between S_m and $S - S_m$ is now considered in defining the heterogeneity of S_m.

(viii) $\mathrm{count}\,a(S_m, S - S_m)$; the sizes of the sets S_m and $S - S_m$ directly influence the number of comparisons that can be made; this raw count can be normalized as in (ix).

(ix) $[1/(n_m(n_m - 1)2n_m(n - n_m))]\mathrm{count}\,a(S_m, S - S_m)$; the number of inconsistencies in the order relation between proximities for object pairs within a subset versus object pairs defined between a subset and its complement is normalized by the number of such possible comparisons to generate the proportion of inconsistencies.

(x) count $b(S_m, S - S_m)$; in contrast to (viii), inconsistent order relations are now counted only between object pairs within S_m and object pairs between S_m and $S - S_m$ when the latter pairs also include one object in common with the object pair within S_m.

(xi) $[1/(2n_m(n_m - 1)(n - n_m))]$count $b(S_m, S - S_m)$; this is a normalization of (x) by the total number of order comparisons made.

(xii) span(S_m); if the sum of the lengths of the minimum spanning trees within subsets is minimized, the same partition is identified as when using the single-link hierarchical clustering method and then choosing that level of the hierarchy with M subsets (compare with the comment made in (v) on min(S_m) and consider minimizing the maximum heterogeneity for this measure).

A numerical illustration. To give a sense of the variability in the clustering results that can be generated by the choice of different definitions for subset heterogeneity and the two optimality criteria of minimizing either the sum of the heterogeneity indices over subsets or their maximum, each of the twelve definitions of subset heterogeneity listed above was matched with the two optimization criteria on the digit proximity data of Table 1.1, from Shepard, Kilpatric, and Cunningham (1975). The ten different optimal partitions corresponding to prechosen $M = 4$ subsets are given below along with a designation of which heterogeneity measure and optimization criterion gave a particular partition and the optimal values achieved in each instance (we use the Roman numeral designation for the heterogeneity measure along with a label of 'sum' or 'max' for the optimality criterion).

The partition in (a), identified as optimal most often, has the clearest interpretation using the structural properties of the digits, i.e., subsets correspond to the additive/multiplicative identities ({0,1}), multiples of 2 ({2,4,8}), multiples of 3 ({3,6,9}), and two odd numbers that are not multiples of 3 ({5,7}). The partitions in (b), (c), (d), and (e) are all subdivisions according to numerical magnitude with each class defined by a consecutive set of digits. The two partitions in (f) and (g) combine (in different ways) the structural and magnitude characteristics of the digits; (h), (i), and (j) all involve one large subset that contains seven objects with three other objects split off into separate subsets. Obviously, the choice of heterogeneity measure and optimization criterion influences which partitions are identified, and although each *is* optimal according to the choices made, there are clearly differing patterns in the proximities being used to achieve this optimality.

Optimal partitions into $M = 4$ subsets, obtained using DPCL1U:

(a) {0,1},{2,4,8},{3,6,9},{5,7}: (i) sum (4.704); (i) max (1.718);
(ii) sum (.921); (ii) max (.286);
(viii) sum (24.0); (viii) max (12.0);
(ix) sum (.235); (ix) max (.125);
(x) sum (12.0); (x) max (6.0);
(xi) max (.214)

3.1. PARTITIONING   21

  (b) {0},{1,2,3,4},{5},{6,7,8,9}: (iii) max (.386)

  (c) {0,1},{2,3,4},{5},{6,7,8,9}: (vi) max (-.224)

  (d) {0},{1},{2},{3,4,5,6,7,8,9}: (xi) sum (.214)

  (e) {0,1},{2,3,4},{5,6,7},{8,9}: (iv) max (.421)

  (f) {0,1},{2,4,8},{3,5,6,9},{7}: (vii) max (-.571)

  (g) {0,1},{2,4,5},{3,6,9},{7,8}: (xii) max (.559)

  (h) {0},{1},{2,3,4,5,6,8,9},{7}: (iii) sum (.443);   (vi) sum (-2.069); (vii) sum (-2.512)

  (i) {0},{1,3,4,5,6,7,9},{2},{8}: (iv) sum (.758)

  (j) {0},{1,2,3,4,6,8,9},{5},{7}: (v) sum (.346);   (v) max (.346); (xii) sum (1.494)

3.1.1 Admissibility Restrictions on Partitions

For each subset heterogeneity measure that could be chosen to characterize optimality, there is an easy extension that can be used to limit the type of subset considered admissible in forming an optimal partition. For example, if there is a spatial context underlying the objects in S, one might wish to impose some type of geographic contiguity constraint so clusters (or subsets) could not be part of an optimal partition, unless they consist of objects that are contiguous in some well-defined sense (e.g., all objects contained in a cluster are within a certain distance of each other, or all objects must be connected through paths defined over adjacent boundaries; see Arabie and Hubert (1996), pp. 15–16, for an overview of this topic). No matter how the admissibility of a cluster might be defined, either through the information present in the proximity matrix $\mathbf{P}$ itself (e.g., the imposition of absolute maximal diameter restrictions), or through some criteria apart from $\mathbf{P}$ (e.g., maximal or minimal subset sizes, geographic contiguity), there is a simple device for avoiding partitions containing inadmissible clusters. One simply defines the heterogeneity of any inadmissible cluster to be a *very* large number. The recursive process remains unchanged, but no partition could be selected as optimal if it contained an inadmissible subset.

As discussed in the previous section, the identification of an optimal partition of S into M classes with a specialization of the GDPP involved an implicit search over *all* possible M-class partitions. This statement is still applicable even when the heterogeneity measure itself may have been augmented to eliminate inadmissible clusters, irrespective of how they might be defined, from appearing in an optimal solution. In some cases, however, it may be possible to use an admissibility criterion to advantage by redefining and reducing the size of the sets $\Omega_1, \ldots, \Omega_K$ needed in carrying out the recursive process, making it possible to approach much larger object set sizes effectively.

A specific restriction discussed at some length in the literature (e.g., Fisher (1958); Rao (1971); Hartigan (1975), Chapter 5; Späth (1980), pp. 61–64) is when there is an assumed object ordering along a continuum that can be taken without loss of generality as $O_1 \prec O_2 \prec \cdots \prec O_n$, and the only admissible clusters are those for which the objects in the cluster form a consecutive sequence or segment. Thus, an optimal partition will consist of M clusters, each of which defines a consecutive segment along the given object ordering.

To tailor the GDPP given in (2.4) and (2.6) to a consecutive-ordering admissibility criterion, each of the sets $\Omega_1, \ldots, \Omega_K$ is now defined by the n subsets of S that contain the objects $\{O_1, \ldots, O_i\}$ for $1 \leq i \leq n$; $\mathcal{F}(A_k)$ is the optimal value for a partitioning of A_k into k classes; a transformation of an entity in Ω_{k-1} (say, A_{k-1}) to one in Ω_k (say, A_k) is possible if $A_{k-1} \subset A_k$; and the cost of the transition is $H(A_k - A_{k-1})$, where $A_k - A_{k-1}$ must contain a consecutive sequence of objects. Again, $\mathcal{F}(A_K)$ for $A_K = S$ defines an optimal solution that can be identified, as always, by working backwards through the recursive process. Similarly, $\mathcal{F}(A_{K-1}), \ldots, \mathcal{F}(A_2)$ for $A_{K-1} = \cdots = A_2 = S$ allows the identification of optimal solutions for $K - 1, \ldots, 2$ classes.

The selection of some prechosen ordering that constrains admissible clusters in a partition obviously does not lead necessarily to the same unconstrained optimal partitions, even though the identical subset heterogeneity measure and optimization criterion are being used. There are, however, several special instances where the original proximity matrix $\mathbf{P}$ is appropriately defined and/or patterned so that the imposition of a particular order constraint does invariably lead to partitions that would also be optimal even when no such order constraint was imposed. One such result dates back to Fisher (1958), who showed that when proximities are squared differences between the values on some (unidimensional) variable, and the order constraint is derived from the ordering of the objects on this variable, then selecting the subset heterogeneity measure we denoted as (iii) (i.e., $[1/(2n_m)]\text{sum}(S_m)$), and minimizing the sum as an optimization criterion, leads to partitions that are not only optimal under the order constraint but are also optimal when unconstrained, i.e., an unconstrained optimal partition will include only those subsets defined by objects consecutive in the given order. (The subset heterogeneity measure in this unidimensional case reduces to the sum of squared deviations of the univariate values for the objects from their mean value within the subset.)

A more general result appears in Batagelj, Korenjak-Černe, and Klavžar (1994), who discuss the special case of when a proximity matrix $\mathbf{P}$ can be row- and column-reordered to display an anti-Robinson form (i.e., the entries within each row and column of $\mathbf{P}$ never decrease when one moves away from a main diagonal entry in any direction). For certain subset heterogeneity measures and optimization criteria, the imposition of the order constraint that displays the anti-Robinson pattern in the row and column reordered proximity matrix leads to partitions that are also optimal when unconstrained. For example, one such case shown by Batagelj, Korenjak-Černe, and Klavžar (1994) would be the use of the sum or the maximum of the subset diameters (the measure in (iv));

3.1. PARTITIONING

other configurations, however, may sometimes fail, e.g., the use of the sum or the maximum of the sum of the proximities within subsets (the measure in (i)).[13]

The choice of an ordering that can be imposed to constrain the search domain for optimal partitions could be directly tied to the task of finding an (optimal) sequencing of the objects along a continuum (which is discussed extensively in Chapter 4). Somewhat more generally, one possible data analysis strategy for seeking partitions as close to optimal as possible would be to construct a preliminary object ordering through some initial optimization process, and possibly one based on another analysis method that could then constrain the domain of search for an optimal partition. Obviously, if one were successful in generating an appropriate object ordering, partitions that would be optimal when constrained would also be optimal unconstrained. The obvious key here is to have some mechanism for identifying an appropriate order to give this possible equivalence (between an optimal constrained partition and one that is optimal unconstrained) a chance to succeed.

As one explicit example of how such a process might be developed for constructing partitions based on an empirically generated ordering for the objects, a recent paper by Alpert and Kahng (1995) proposed a three-stage process. First, the objects to be partitioned are embedded in a Euclidean representation with a specific multidimensional scaling strategy (Alpert and Kahng (1995) suggest a method they attribute to Hall (1970), but that was actually developed much earlier by Guttman (1968), who used it to develop an initial spatial configuration for the objects in his approach to nonmetric multidimensional scaling). Second, by heuristic methods, a path among the n objects in the Euclidean representation is identified (with hopefully close to minimal length) and used to define a prior ordering for the objects and to constrain the subsets that would be present in a partition. Third, a DP strategy of the same general form we have described is carried out to obtain a partitioning of S. (Alpert and Kahng (1995) apparently believe the DP recursion they suggest is new to the literature but except for some trivial differences, it is identical in form to that first suggested by Fisher (1958), which we have summarized above.)[14]

A numerical illustration. To show the wide variability that the choice of heterogeneity measure and optimization criterion may produce in the identification of now constrained optimal partitions, the same numerical example of the last section was replicated, but with the order constraint imposed that the subsets must be consecutive in their actual digit magnitudes. Twelve different optimal partitions with $M = 4$ subsets were obtained and are given below with the same designations as in the previous example. Those partitions labeled as (b), (c), (d), and (e) correspond to the same labels in the unconstrained illustration, where partitions consistent with digit magnitude were identified. These were obtained, as expected, using the heterogeneity measures and optimization criteria that led to their previous identification (they were also found here for several other choices as well). To provide a simple numerical check, as can be seen by comparing the actual optimal values given below for the constrained partitions and those (given previously) for the unconstrained partitions, whenever a specific heterogeneity measure/optimization criterion led to an unconstrained optimal partition in which subsets were not consecutive in digit magnitude, the constrained minimal values were larger (as they are required to be).

Optimal (linear order restricted) partitions into $M = 4$ subsets, obtained using DPCL1R:

(a) {0},{1,2,3},{4,5},{6,7,8,9}: (x) sum (50.0)

(b) {0},{1,2,3,4},{5},{6,7,8,9}: (iii) max (.386);   (ix) max (.215);
(xii) sum (1.735)

(c) {0,1},{2,3,4},{5},{6,7,8,9}: (ii) sum (1.0643);
(vi) max (-.224)

(d) {0},{1},{2},{3,4,5,6,7,8,9}: (ix) sum (.243);   (xi) sum (.214);
(v) sum (.400)

(e) {0,1},{2,3,4},{5,6,7},{8,9}: (ii) max (.404);   (i) sum (5.704);
(i) max (2.426);   (iv) max (.421);
(viii) sum (136.0)

(f) {0,1,2,3},{4},{5},{6,7,8,9}: (xi) max (.208)

(g) {0},{1,2,3,4,5,6},{7},{8,9}: (vii) sum (-2.474); (v) max (.396)

(h) {0},{1},{2,3,4,5,6},{7,8,9}: (vii) max (-.565)

(i) {0,1},{2,3,4},{5,6},{7,8,9}: (xii) max (.792)

(j) {0},{1,2,3,4,5,6,7},{8},{9}: (iv) sum (.796)

(k) {0},{1},{2,3,4,5,6,7,8},{9}: (iii) sum (.458);   (vi) sum (-2.000)

(l) {0,1,2},{3,4},{5,6,7},{8,9}: (viii) max (56.0);   (x) max (18.0)

There are two generalizations of the use of a single linear object ordering to define subset admissibility that are worth mentioning, although we will not pursue these extensions in any detail here. First, if the object set can be partitioned into, say, $T > 1$ subsets, where there is a consecutive object ordering that must be satisfied within each such subset, then it is again possible to reduce the number of entities present in $\Omega_1, \ldots, \Omega_K$ (although obviously not as much as when $T = 1$). For example, if $T = 2$, the subsets of S that must be present in $\Omega_k, 1 \leq k \leq K$, are those that contain the first i and i' consecutive objects within each given ordering, including those subsets that may contain no objects from one of the orderings. Similar extensions exist for $T > 2$, but clearly the limiting case when $T = n$ would move us all the way back to where the sets $\Omega_k, 1 \leq k \leq K$, must include all nonempty subsets of S, i.e., to the unconstrained M-class partitioning task of Section 3.1. Second, the notion of subset admissibility based on a single linear object ordering can be extended to the use of a fixed *circular* ordering that could be taken without loss of generality as $\cdots \prec O_1 \prec O_2 \prec \cdots \prec O_n \prec O_1 \prec \cdots$. Thus, an optimal partition would consist of M clusters, all of which would be segments in the circular object ordering. Each of the sets $\Omega_1, \ldots, \Omega_K$ would now be defined by the n^2 subsets of S that contain the objects $\{O_k, \ldots, O_i\}$ for $k = 1, \ldots, n; i = k, \ldots, n, 1, \ldots, k-1$. The optimal value $\mathcal{F}(A_k)$ is associated with a partitioning of $A_k \in \Omega_k$ into k classes; a transformation of an entity in Ω_{k-1} (say, A_{k-1}) to one in Ω_k (say, A_k) is possible if $A_{k-1} \subset A_k$ and $A_k - A_{k-1}$ is a consecutive sequence of objects with respect to the fixed circular order. Again, the cost of transition is $H(A_k - A_{k-1})$. An optimal solution can be obtained from $\min_k \mathcal{F}(\{O_k, \ldots, O_n, O_1, \ldots, O_{k-1}\})$, where $\{O_k, \ldots, O_{k-1}\} \in \Omega_K$ for $1 \leq k \leq n$, and then working backwards through the recursive process. Analogously, an optimal solution into $K - h$ subsets for $1 \leq h \leq K - 2$ could be identified from $\min_k \mathcal{F}(\{O_k, \ldots, O_{k-1}\})$, where $\{O_k, \ldots, O_{k-1}\} \in \Omega_{K-h}$. A further discussion of this type of circular ordering restriction is given in Alpert and Kahng (1997) in the context of defining the circular ordering restriction by some secondary method of analysis. If we so wished, a single circular object ordering could be generalized to multiple circular object orderings analogous to the use of multiple linear object orderings noted above.

3.1.2 Partitioning Based on Two-Mode Proximity Matrices

As noted in the introduction to Section 3, one possible extension of the optimal partitioning task, whether unconstrained or constrained by some given object order, is to the context where S itself is the union of two disjoint sets, say $S = S_A \cup S_B$, and the only proximities available are defined from the objects in S_A to those in S_B (see endnote 7). If it is assumed that S_A and S_B each contain n_A and n_B objects, respectively, then a more convenient notation for the available data might be as a matrix of order $n_A \times n_B$, which we denote by $\mathbf{Q} = \{q_{ij}\}$. The rows and columns of $\mathbf{Q}$ specify two modes for the data (to use the terminology from Tucker (1964)), and q_{ij} is now the proximity (assumed to be keyed as a dissimilarity) between the ith object in S_A and the jth object in S_B. For convenience of reference, the objects in S_A and S_B, respectively, will now be denoted by $\{r_1, \ldots, r_{n_A}\}$ and $\{c_1, \ldots, c_{n_B}\}$ (intuitively referring to the row and column objects). (We might also note at the outset that both S_A and S_B *may* refer to the same set of objects, where the original proximity matrix between the objects from this common set could be nonsymmetric [and even have meaningful entries along the main diagonal]. Thus, for convenience, the row and column objects are treated as if they were distinct, and we go on to consider the task as one of partitioning a combined row and column object set.)

One direct approach to partitioning based on such two-mode data merely assumes we have the single set S ($= S_A \cup S_B$) and generalizes all the heterogeneity measures introduced earlier to ignore missing proximities within S_A or within S_B (and with the appropriate adjustments in taking averages for only the number of proximities summed, and so on), to make any subset of S inadmissible if it contains only objects from S_A or from S_B.[15] Because of the constraint that all subsets must contain objects from both S_A and S_B, the maximum number of subsets in any partition of $S_A \cup S_B$ will be the minimum of n_A and n_B.[16]

Although a comparable extension to order-constrained partitioning based on **Q** could also be carried out using a single joint ordering of the row and column objects, a more general possibility is to allow separate row and column orders to restrict the desired clusters. Thus, the aim would be to provide optimal M-class partitions of $S = S_A \cup S_B$, based on some heterogeneity measure that uses only the proximities in **Q**, and minimize either the sum or the maximum of the heterogeneity measures over the subsets in the partition. Each class in an optimal partition is constrained to contain objects from both S_A and S_B consecutive in their respective row or column orderings (that without loss of any generality can be taken to be $r_1 \prec \cdots \prec r_{n_A}$ and $c_1 \prec \cdots \prec c_{n_B}$). The storage requirements needed to implement the GDPP and the use of separate row and column ordering restrictions will be greater than for a single joint ordering because many more entities have to be defined through the sets $\Omega_1, \ldots, \Omega_K$.[17]

To show explicitly how the GDPP in (2.4) and (2.6) can be tailored to encompass consecutive order admissibility criteria separately on the rows and columns, each of the sets $\Omega_1, \ldots, \Omega_K$ would now be defined by the $n_A n_B$ subsets of $S_A \cup S_B$ that contain the objects $\{r_1, \ldots, r_i, c_1, \ldots, c_j\}$ for $1 \leq i \leq n_A$, $1 \leq j \leq n_B$; $\mathcal{F}(A_k)$ for $A_k \in \Omega_k$ is the optimal value for a partitioning of A_k into k classes. A transformation of an entity in Ω_{k-1} (say, A_{k-1}) to one in Ω_k (say, A_k) has a cost of $H(A_k - A_{k-1})$, and is possible if (a) $A_{k-1} \subset A_k$; (b) $(A_k - A_{k-1}) \cap S_A \neq \emptyset$ and forms a consecutive set of objects in the given row order; and (c) $(A_k - A_{k-1}) \cap S_B \neq \emptyset$ and forms a consecutive set of objects in the given column order. An optimal solution is given by $\mathcal{F}(A_K)$ for $A_K = S_A \cup S_B$, and an actual solution can be constructed by the usual process of working backwards though the steps of the recursion. Also, $\mathcal{F}(A_{K-1}), \ldots, \mathcal{F}(A_2)$ for $A_{K-1} = \cdots = A_2 = S_A \cup S_B$ provide the optimal values for partitions containing from $K-1$ to 2 classes.

A numerical illustration. To provide an illustration of the results obtainable in optimally partitioning a two-mode proximity matrix, we use the 11×9 matrix in Table 1.3 giving the proximities between eleven goldfish retinal receptors (the row objects) and nine wavelengths (the column objects). Selecting the subset heterogeneity measure defined by the subset diameter (iv) and minimizing the maximum diameter over the classes of a partition produced the following optimal partitions for two through six classes (the nine digits that correspond to the wavelengths are denoted by underlining):

3.1. PARTITIONING

Optimal partitions into two through six classes, obtained using DPCL2U:

Number of Classes	Partition	Maximum Diameter
2	{1,5,7,10,11,<u>1</u>,<u>2</u>,<u>3</u>,<u>7</u>,<u>8</u>}, {2,3,4,6,8,9,<u>4</u>,<u>5</u>,<u>6</u>,<u>9</u>}	155.0
3	{1,4,5,10,11,<u>2</u>,<u>8</u>}, {2,3,6,8,9,<u>4</u>,<u>6</u>,<u>9</u>}, {7,<u>1</u>,<u>3</u>,<u>5</u>,<u>7</u>}	127.0
4	{1,4,5,6,7,10,<u>2</u>}, {2,<u>1</u>,<u>5</u>,<u>6</u>,<u>7</u>}, {3,8,9,<u>4</u>,<u>9</u>}, {11,<u>3</u>,<u>8</u>}	91.0
5	{1,5,7,11,<u>2</u>}, {2,<u>1</u>,<u>5</u>,<u>6</u>}, {3,8,9,<u>4</u>,<u>9</u>}, {4,6,<u>7</u>}, {10,<u>3</u>,<u>8</u>}	78.0
6	{1,5,7,11,<u>2</u>}, {2,<u>1</u>,<u>5</u>}, {3,9,<u>4</u>,<u>9</u>}, {4,6,<u>7</u>}, {8,<u>6</u>}, {10,<u>3</u>,<u>8</u>}	75.0

The nine wavelengths are ordered from longest (#<u>3</u>, red) to shortest (#<u>9</u>, violet) as <u>3</u> → <u>8</u> → <u>2</u> → <u>7</u> → <u>1</u> → <u>5</u> → <u>6</u> → <u>4</u> → <u>9</u>. Except for one anomaly in the placement of <u>3</u> for the optimal partition into three classes (which might be viewed as too crude a partition, given the size of the maximum diameter over the three subsets compared with that for the optimal partitions containing four, five, and six subsets), all the optimal partitions consistently group consecutive wavelengths with a subset of the receptors, i.e., specific receptors appear differentially sensitive to specific wavelength ranges.

Continuing this example, we will impose row and column orders identified later in Section 4.1.3 in an optimal joint sequencing of the rows and columns along a continuum:

row order: 10 → 11 → 5 → 7 → 4 → 6 → 2 → 8 → 9 → 3;

column order: <u>3</u> → <u>8</u> → <u>2</u> → <u>7</u> → <u>1</u> → <u>5</u> → <u>6</u> → <u>4</u> → <u>9</u> (this column order is consistent with decreasing wavelength).

Based on these restrictions, the optimal partitions into two through six classes are given below:

Optimal (constrained) partitions into two through six classes, obtained using DPCL2R:

Number of Classes	Partition	Maximum Diameter
2	{1,5,7,10,11,<u>1</u>,<u>2</u>,<u>3</u>,<u>7</u>,<u>8</u>}, {2,3,4,6,8,9,<u>4</u>,<u>5</u>,<u>6</u>,<u>9</u>}	155.0
3	{1,4,<u>1</u>,<u>2</u>,<u>7</u>}, {2,3,6,8,9,<u>4</u>,<u>5</u>,<u>6</u>,<u>9</u>}, {5,7,10,11,<u>3</u>,<u>8</u>}	143.0
4	{1,4,5,6,7,11,<u>2</u>}, {2,<u>1</u>,<u>5</u>,<u>6</u>,<u>7</u>}, {3,8,9,<u>4</u>,<u>9</u>}, {10,<u>3</u>,<u>8</u>}	91.0
5	{1,5,7,11,<u>2</u>}, {2,<u>1</u>,<u>5</u>,<u>6</u>}, {3,8,9,<u>4</u>,<u>9</u>}, {4,6,<u>7</u>}, {10,<u>3</u>,<u>8</u>}	78.0
6	{1,5,7,11,<u>2</u>}, {2,<u>1</u>,<u>5</u>}, {3,9,<u>4</u>,<u>9</u>}, {4,6,<u>7</u>}, {8,<u>6</u>}, {10,<u>3</u>,<u>8</u>}	75.0

The optimal constrained partitions into two, five, and six subsets are the same as those for the unconstrained results. For the optimal four-class partition—although the optimal diameter values are the same (i.e., a value of 91.0)—receptors 10 and 11 are interchanged between the constrained and unconstrained results. In the restricted four-class partition, the consecutive row order that was imposed forces such an interchange. The optimal constrained three-class partition is not the same as the unconstrained three-class partition. It has a large maximum diameter of 143.0, reflecting again the possibly too crude grouping that an optimal three-class partition would impose on these data (as observed in the unconstrained optimal three-class partition and the anomalous placement of wavelength 3).

The discussion in this section has emphasized two-mode proximity data defined between the distinct sets S_A and S_B, where subset heterogeneity is some function of the proximities attached to the 2-tuples within a subset (and/or the complement), with each such 2-tuple containing an object from each of the two modes. Without pursuing the topic, we will note here that there are very direct generalizations of the partitioning task for t-mode data (for $t > 2$) and for proximities defined for each t-tuple of objects. Subsets would now contain at least one object from each mode, and subset heterogeneity would be constructed by some function of the proximities attached to the t-tuples within each subset and/or the complement. Also, it would be possible to develop specializations of the GDPP in this general t-mode context (and for $t = 2$ as well) that would incorporate specific order constraints for some modes but not others. This differential set of order constraints would characterize the admissibility of some of the partitions and could be adopted for reducing the size of the sets $\Omega_1, \ldots, \Omega_K$ needed in carrying out the appropriate recursive process.

3.2 Hierarchical Clustering

In contrast to the partitioning task in which a single collection of M mutually exclusive and exhaustive subsets is sought, the problem of hierarchical clustering will be characterized by the search for an optimal collection of partitions of S, which for now we denote generically as $\mathcal{P}_1, \mathcal{P}_2, \ldots, \mathcal{P}_n$. Here, $\mathcal{P}_1$ is the (trivial) partition where all n objects from S are placed into n separate classes, $\mathcal{P}_n$ is the (also trivial) partition where a single subset contains all n objects, and $\mathcal{P}_k$ is obtained from $\mathcal{P}_{k-1}$ by uniting some pair of classes present in $\mathcal{P}_{k-1}$. (Given several generalizations encountered later, and because for now only a single pair of classes will be united in $\mathcal{P}_{k-1}$ to form $\mathcal{P}_k$, we will typically employ the adjective of 'full' to refer to the partition hierarchy $\mathcal{P}_1, \ldots, \mathcal{P}_n$.) Obviously, in moving from the identification of single partitions (as in Section 3.1) to the construction of a full partition hierarchy, we also have traveled dramatically up the combinatorial ladder according to the number of such structures to be considered by any partial enumeration method. It should therefore not be surprising that (a) in comparison with the generation of single optimal partitions and (b) with no further constraints on the optimization task, only smaller object

3.2. HIERARCHICAL CLUSTERING

sets can be effectively approached with whatever GDPP specializations might be proposed for the construction of an optimal partition hierarchy (e.g., given current storage resources, the effective limit for object set sizes is, say, in the lower teens).

Given the broadly stated task of identifying an optimal partition hierarchy, there are still many variations possible on how the form of the optimality criterion might be defined. One very flexible option includes several alternatives as special cases, and concentrates directly on minimizing the sum of transition costs, however they might be defined, between successive partitions in a hierarchy. Specifically, suppose we let $T(\mathcal{P}_{k-1}, \mathcal{P}_k)$ denote some measure of transition cost between two partitions $\mathcal{P}_{k-1}$ and $\mathcal{P}_k$, where $\mathcal{P}_k$ is constructed from $\mathcal{P}_{k-1}$ by uniting two classes in the latter partition. An optimal full partition hierarchy $\mathcal{P}_1, \ldots, \mathcal{P}_n$ is one for which the sum of the transition costs,

$$\sum_{k>2} T(\mathcal{P}_{k-1}, \mathcal{P}_k), \qquad (3.1)$$

is minimized.

There are many ways to operationalize a transition cost between two partitions, but the most obvious measures would involve some direct function of the between- and within-class proximities for the two subsets in $\mathcal{P}_{k-1}$ merged to form the new class in $\mathcal{P}_k$. However, if we so choose, we might also proceed more indirectly. For example, suppose a heterogeneity measure $L(\cdot)$ is first defined that assigns a value to every possible partition (with a cost interpretation, so that smaller values represent 'better' partitions). As an illustration, $L(\mathcal{P}_k)$ might merely be the sum or the maximum of the heterogeneity indices over the subsets constituting $\mathcal{P}_k$, using one of the alternatives noted in Section 3.1. The transition $T(\mathcal{P}_{k-1}, \mathcal{P}_k)$ could then be defined simply as $L(\mathcal{P}_k) - L(\mathcal{P}_{k-1})$, i.e., as the difference between the heterogeneity values attached to the partitions $\mathcal{P}_k$ and $\mathcal{P}_{k-1}$.

Necessarily limiting the scope of our present task so as to provide a direct connection to the partitioning discussion of the previous section, we will restrict the notion of a transition cost between two partitions to the use of a heterogeneity measure that can be attached to the new subset in $\mathcal{P}_k$ formed from the union of two subsets in $\mathcal{P}_{k-1}$. Even more specifically, we will use only those indices of subset heterogeneity previously introduced in Section 3.1. (This limitation will be true except for those instances developed in Section 3.2.1 that used the GDPP to attempt an optimal fitting of an ultrametric to a given proximity matrix.) Moreover, only the optimization criterion defined in (3.1) as a sum will be considered, because a minimization of a maximum transition cost will generally be of more limited utility in identifying a good full partition hierarchy. Explicitly, because subset heterogeneity measures tend to increase as fewer subsets define a partition and the transition cost is given by the heterogeneity of the new subset formed in a partition, the use of an optimization criterion based solely on minimizing the maximum transition cost would typically overemphasize the single partition in a hierarchy that contains only two subsets, at the expense of the hierarchy below that point. In short, then, our discussion will generally be limited to the identification of optimal full partition hierarchies that minimize the sum of the heterogeneity measures for the $n-2$ (new) subsets formed in progressing from $\mathcal{P}_1$ to $\mathcal{P}_n$.

The application of the GDPP in (2.4) to the minimization of (3.1), apart from any specific definition of transition cost, would first define n sets, $\Omega_1, \ldots, \Omega_n$, where Ω_k contains all partitions of the n objects in S into $n-k+1$ classes. The value $\mathcal{F}(A_k)$ for $A_k \in \Omega_k$ is the optimal sum of transition costs up to the partition A_k; a transformation of an entity in Ω_{k-1} (say, A_{k-1}) to one in Ω_k (say, A_k) is possible if A_k is obtainable from A_{k-1} by uniting two classes in A_{k-1}, and has cost $C(A_{k-1}, A_k) \equiv T(A_{k-1}, A_k)$. Beginning with an assumed value for $\mathcal{F}(A_1)$ of 0 for the single entity $A_1 \in \Omega_1$ (which is the partition of S into n subsets each containing a single object), and constructing $\mathcal{F}(A_k)$ recursively for $2, \ldots, n$, an optimal solution is identified by $\mathcal{F}(A_n)$ for the single entity $A_n \in \Omega_n$ defined by the partition containing all n objects in a single class. A partition hierarchy attaining this optimal value is again obtained by working backward through the recursion.

Given the definition of the sets $\Omega_1, \ldots, \Omega_n$, and because the recursive strategy progresses from Ω_1 to Ω_n guided by some transition measure $T(\mathcal{P}_{k-1}, \mathcal{P}_k)$ and by the total cost measure in (3.1), we might use the term 'agglomerative' for this approach because pairs of sets are successively united to produce the next partition in the sequence. The well-known heuristic agglomerative strategies for hierarchical clustering can also be interpreted as being based on measures of transition between partitions, but the process used in obtaining the partition hierarchy is a 'greedy' one in which starting at $\mathcal{P}_1$, each partition $\mathcal{P}_k$ is obtained in turn from $\mathcal{P}_{k-1}$ by merely choosing the one with the lowest transition cost, and continuing until $\mathcal{P}_n$ is reached. Thus, in general there is no guarantee that such a greedy heuristic will minimize the sum in (3.1), except for one definition of $T(\mathcal{P}_{k-1}, \mathcal{P}_k)$ that characterizes the single-link hierarchical clustering method. If $T(\mathcal{P}_{k-1}, \mathcal{P}_k)$ is defined to be a minimum proximity value between the pair of subsets in $\mathcal{P}_{k-1}$ united to form the new set in $\mathcal{P}_k$ (or equivalently, the heterogeneity value defined in (v) for the new subset), and $\mathcal{P}_k$ is formed from $\mathcal{P}_{k-1}$ by uniting the associated pair of subsets, then the index in (3.1) is minimized by the greedy heuristic. This result stems from the connections that exist between single-link hierarchical clustering and the construction of minimum spanning trees within a weighted complete graph based on the proximity matrix $\mathbf{P}$ (e.g., see Barthélemy and Guénoche, 1991, Chapter 3). It would be very computationally inefficient, to say the least, to rely on the GDPP to construct a single-link partition hierarchy, given that a simple greedy heuristic is sufficient, but the theoretical possibility still exists.

Before providing a few numerical illustrations of the results obtainable in constructing optimal partition hierarchies based on optimizing the sum of heterogeneity measures over the $n-2$ subsets formed while proceeding from $\mathcal{P}_1$ to $\mathcal{P}_n$, it may be helpful first to clarify several related points about the conduct of the optimization process itself. First, for any of the subset heterogeneity measures introduced in Section 3.1 based solely on the proximities within a particular cluster (i.e., those defined in (i), (ii), (iii), (iv), (v), and (xii)), it can be shown (although we will not digress here to do so formally) that in any optimal partition hierarchy, the heterogeneity measures attached to the $n-2$ new subsets constructed in the process must be monotonic with respect to subset inclusion. Explicitly, if S_a and S_b denote two of the $n-2$ subsets identified while constructing the partition hierarchy, and if $S_a \subset S_b$, then $H(S_a) \leq H(S_b)$. Thus, to exploit this result in gaining some computational advantage when carrying out the recursive process of generating $\mathcal{F}(A_k)$ from $\mathcal{F}(A_{k-1})$, and if one of the subset heterogeneity indices noted above is being used, then only those transitions from a partition A_{k-1} to A_k are allowed for which the heterogeneity of the new subset formed is at least as large as the heterogeneity values for all the subsets defining the partition A_{k-1}.

For a subset heterogeneity measure that includes some function of the proximities between the subset and its complement (i.e., those alternatives other than (i), (ii), (iii), (iv), (v), and (xii)), the monotonicity requirement is not necessarily present in an optimal partition hierarchy and, therefore, an attempt to impose it could lead to a result that would be suboptimal according to the sum of the heterogeneity measures over the $n-2$ subsets identified in the process of hierarchy generation. From a broader perspective on the allowability of certain transitions between partitions in Ω_{k-1} and Ω_k, it may be of interest, much as in our discussion of constraining the construction of single partitions, to impose other admissibility criteria and permit only certain transitions that produce new subsets satisfying some particular constraint (e.g., geographic contiguity, consecutiveness with respect to a particular object ordering, and so on). (We will discuss a version of constrained hierarchical clustering in Section 3.2.2 based on an assumed ordering for the objects in S, and in Section 3.2.1 within the context of using the GDPP to attempt an optimal least-squares fitting of an ultrametric to a given proximity matrix.)

As a second point, we note that the storage requirements for carrying out a GDPP approach to hierarchical clustering are rather enormous, and in effect, this drawback is the major determinant of how large the object sets that might be approached through the GDPP can be.[18] The computational efforts needed in working through the recursive process are also great, however, and for at least some subset heterogeneity measures (in fact, for all our exemplars except (vi) and (vii) that include the use of the average proximity between a subset and its complement), it may be possible to reduce this effort substantially with an upper bound (separately obtained) on the value that should be achievable for the optimization criterion for an optimal partition hierarchy. Specifically, for each of the subset heterogeneity measures we have been considering (except, as noted above, for (vi) and (vii)), the increment to the total optimization criterion (defined by the sum of the heterogeneity measures over the $n-2$ subsets formed in any partition hierarchy) must be nonnegative when considering a transition from a partition A_{k-1} to A_k. Thus, if one has a known (upper) bound achievable from another partition hierarchy based simply on the sum of the $n-2$ heterogeneity values for the $n-2$ subsets formed in that hierarchy, transitions between partitions A_{k-1} and A_k can be considered inadmissible and thus ignored whenever the cumulative sum of the heterogeneity values resulting from that transition is already greater than the achievable bound. The upper bound that we will rely on is obtained from a simple application of the greedy heuristic, in which a partition hierarchy is formed by successively choosing the minimal value over all possible subsets that could be formed at each successive level of a partition hierarchy starting with $\mathcal{P}_1$ and progressing step-by-step to $\mathcal{P}_n$ by uniting subsets that achieve this minimal value at each stage. In the numerical examples to follow, the partition hierarchy obtained by an application of the greedy strategy will be given, as well as the bound obtained and subsequently used to reduce the computational efforts in identifying an optimal solution.

The recursive process of constructing a partition hierarchy emphasizes obtaining a full hierarchy that progresses invariably from $\mathcal{P}_1$, in which all objects are in separate classes, to $\mathcal{P}_n$, in which all objects are placed into one class. However, the information generated within this recursion can also identify a partial partition hierarchy, $\mathcal{P}_1, \ldots, \mathcal{P}_k$, where $\mathcal{P}_k$ contains $n - k + 1$ subsets, which is optimal with respect to the same sum of heterogeneity measures for the $k - 1$ subsets formed up through $\mathcal{P}_k$. The identification of the minimum value for $\mathcal{F}(A_k)$ over all $A_k \in \Omega_k$ is required, and then one works backwards from that point in the recursion to identify the optimal partial partition hierarchy.[19]

Numerical illustrations. Given the different subset heterogeneity measures introduced in Section 3.1, the number of illustrations that would use just these for constructing the partition hierarchy is still large. Thus, in the interests of brevity, we will only discuss the two measures based on the diameter of a subset (the index in (iv)) and the sum of proximities within a subset (the index in (i)). We begin with the latter.

The optimal partition hierarchy for the digit proximity data of Table 1.1 that minimizes the sum of the eight subset heterogeneity measures (each defined by the sum of within-cluster proximities) obtained in progressing from $\mathcal{P}_1$ to $\mathcal{P}_{10}$ is as follows (in providing a partition hierarchy here and later, single object classes are generally not listed at any level):

An optimal partition hierarchy based on the sum of within-cluster proximities, obtained using DPHI1U:

Level	Partition	Index of New Subset	Cumulative Sum
1	(all digits separate)	—	—
2	{2,4}	.059	.059
3	{2,4},{3,9}	.263	.322
4	{2,4},{3,9},{5,7}	.400	.722
5	{2,4},{3,9},{5,7},{0,1}	.421	1.143
6	{2,4,8},{3,9},{5,7},{0,1}	.672	1.815
7	{2,4,8},{3,6,9},{5,7},{0,1}	.859	2.674
8	{2,4,8},{3,5,6,7,9},{0,1}	4.144	6.818
9	{0,1,2,4,8},{3,5,6,7,9}	4.903	11.721
10	(all digits together)	—	36.077

Clearly, up to the four subsets present at level 7, the structural properties of

3.2. HIERARCHICAL CLUSTERING

the digits are paramount; after level 7, the subsets that are successively united are apparently determined more from numerical magnitude. In any case, the optimal value of the sum of the heterogeneity values for the eight new subsets is 11.721 (and 36.077 for the full partition hierarchy when the one additional subset is included where all digits are placed together).

Using the same subset heterogeneity measure (as above), the partition hierarchy obtained by the greedy heuristic is given below (as noted earlier, the cumulative value of 13.667 obtained at level 9 for this hierarchy was used as a computational bound in limiting admissible transitions when generating the optimal partition hierarchy).

A partition hierarchy based on the greedy optimization of the sum of within-cluster proximities, obtained using DPHI1U:

Level	Partition	Index of New Subset	Cumulative Sum
1	(all digits separate)	—	—
2	{2,4}	.059	.059
3	{2,4},{3,9}	.263	.322
4	{2,4},{3,9},{6,8}	.350	.672
5	{2,4},{3,9},{6,8},{5,7}	.400	1.072
6	{2,4},{3,9},{6,8},{5,7},{0,1}	.421	1.493
7	{2,4,6,8},{3,9},{5,7},{0,1}	1.831	3.324
8	{2,4,6,8},{3,5,7,9},{0,1}	2.735	6.059
9	{0,1,2,4,6,8},{3,5,7,9}	7.608	13.667
10	(all digits together)	—	38.023

Obviously, the greedy heuristic did not produce a full partition hierarchy that was optimal with respect to the sum of the heterogeneity measures over all eight new subsets formed (i.e., a value of 13.667 was achieved for such a sum, compared to 11.721 for the optimal hierarchy). However, it is interesting to note that the cumulative sum of 6.059 within the greedy hierarchy up to level 8 (which includes three subsets) is smaller than the cumulative sum to that point of 6.818 for the optimal hierarchy.[20] Moreover, this level 8 partition in the greedy hierarchy can be given a much clearer interpretation than its counterpart for the optimal hierarchy, i.e., the level 8 greedy partition of {{0,1},{2,4,6,8},{3,5,7,9}} is a split of the digits into the additive/multiplicative identities and subsets of the even and odd digits. By concentrating on the global optimality of the full hierarchy, this latter partition was obviously not identified in the process.

The ambiguities inherent in comparing the hierarchies obtained with the optimal and greedy methods get even more complicated if we inspect an optimal partial partition hierarchy (given below) ending at level 8. The cumulative sum to level 8 of 5.809 is the optimal value that can be achieved for a partial partition hierarchy up to this level. However, the interpretation for the final partition at level 8 must again be made from a mixture of the structural characteristics of the digits and their magnitudes.

An optimal partial partition hierarchy based on the sum of within-cluster proximities, obtained using DPHI1U:

Level	Partition	Index of New Subset	Cumulative Sum
1	(all digits separate)	—	—
2	{2,4}	.059	.059
3	{2,4},{6,9}	.296	.355
4	{2,4},{6,9},{1,3}	.346	.701
5	{2,4},{6,9},{1,3},{5,7}	.400	1.101
6	{2,4,8},{6,9},{1,3},{5,7}	.672	1.773
7	{2,4,8},{6,9},{0,1,3},{5,7}	1.476	3.249
8	{2,4,5,7,8},{6,9},{0,1,3}	2.560	5.809

The use of the subset heterogeneity criterion defined by the subset diameter (the index in (iv)) constructs the optimal full partition hierarchy given below. In contrast to the use of the previous subset heterogeneity measure (defined as the sum of within-cluster proximities), both the optimal full hierarchy and that obtained with a greedy heuristic turn out to be identical. In fact, all optimal partial hierarchies are merely those sections of the optimal full hierarchy up to the appropriate number of subsets (i.e., here, the optimal partial hierarchies are nested). This optimal partition hierarchy is very similar to that given for the subset heterogeneity measure defined by the sum of within-cluster proximities except for the order in which new subsets are formed; e.g., at level 7, the partitions are identical. Generally, the same structural properties of the digits can be used to interpret partitions at levels below 8, and a mixture of digit magnitude and structural properties account for levels 8 and 9.[21]

An optimal partition hierarchy based on the sum of subset diameters, obtained using DPHI1U:

Level	Partition	Index of New Subset	Cumulative Sum
1	(all digits separate)	—	—
2	{2,4}	.059	.059
3	{2,4},{3,9}	.263	.322
4	{2,4},{3,6,9}	.300	.622
5	{2,4,8},{3,6,9}	.367	.989
6	{2,4,8},{3,6,9},{5,7}	.400	1.389
7	{2,4,8},{3,6,9},{5,7},{0,1}	.421	1.810
8	{2,4,8},{3,5,6,7,9},{0,1}	.592	2.402
9	{3,4,5,6,7,8,9},{0,1}	.808	3.210
10	(all digits together)	.909	4.119

Concerning the example just discussed, we might comment that the construction of an optimal partition hierarchy minimizing the sum of the subset diameters solves a very long-standing optimization problem from the classification literature (e.g., see Hubert (1974b), p. 18). Specifically, the set of all possible partition hierarchies can be used to induce an equivalence relation on (i.e., a partitioning of) the collection of ordered spanning trees, where an ordered spanning tree is a spanning tree among the n objects with edges in the tree weighted by the proximities from $\mathbf{P}$, and in which an order is imposed on the successive deletion of edges in the tree. The weight of an ordered spanning tree is merely the sum of the proximities on the $n-1$ edges present in the tree, and two ordered spanning trees are in the same equivalence class if they induce the same partition hierarchy. The partition hierarchy constructed by the single-link method can be obtained through the minimal spanning tree in which edges are deleted according to decreasing weight, and identified in turn with a particular equivalence class of ordered spanning trees.

Within this later equivalence class of ordered spanning trees that all induce the same single-link partition hierarchy, there will be a spanning tree (possibly nonunique) of maximal length; because of the finiteness of each possible equivalence class, a maximal length spanning tree exists within each equivalence class. Thus, single-link hierarchical clustering can be reinterpreted as the selection of a particular equivalence class of ordered spanning trees effected by minimizing the minimal tree over all possible equivalence classes (and with no regard for the maximal spanning tree within its equivalence class). In contrast, the use of the sum of the subset diameters as an optimization criterion can be interpreted as the selection of an equivalence class of ordered spanning trees minimizing the maximal spanning tree (where in such an optimal maximal spanning tree the edges will again be deleted according to decreasing magnitude.)

There are a few additional points to note in concluding this general discussion of using the GDPP for the hierarchical clustering task. The first involves the possibility of carrying out the hierarchical clustering task using the GDPP but for two-mode proximity data. Just as the partitioning task using the GDPP was extended in Section 3.1.2 to an $n_A \times n_B$ two-mode proximity matrix $\mathbf{Q}$ by uniting the distinct object sets forming the rows and columns of $\mathbf{Q}$ and directly generalizing the subset heterogeneity measures to ignore missing data, an exactly analogous extension could be made in the hierarchical clustering context. Although it could be carried out immediately, only very small data sets could be handled by such an extension, i.e., a total number of row and column objects in the lower teens. This limit is probably too small for empirical applications, so most realistic uses for two-mode hierarchical clustering would typically need a heuristic application of the GDPP, which will be discussed in Section 5.1 (but which could provide the optimal hierarchical clustering [of a small two-mode proximity matrix] as a special case).

A second point concerns our choice of how the GDPP for the hierarchical clustering task was developed agglomeratively by defining the sets $\Omega_1, \ldots, \Omega_n$ so that Ω_k contained all partitions of the n objects in S into $n-k+1$ classes. The recursive process started from the trivial partition in Ω_1 and proceeded step-by-step to Ω_n. Instead of proceeding agglomeratively from Ω_1 to Ω_n, we could have chosen a divisive strategy that would reverse the order by starting from Ω_n and moving to Ω_1. Explicitly, we would begin with an assumed value for $\mathcal{F}(A_n)$

of 0 for the single entity $A_n \in \Omega_n$ and define $\mathcal{F}(A_{n-k})$ for $k = 1, \ldots, n-1$ (where $A_{n-k} \in \Omega_{n-k}$) as the optimal value for the sum of transition costs for the partition A_{n-k} consisting of $k+1$ classes. A transformation of an entity in Ω_{n-k} (say, A_{n-k}) to one in Ω_{n-k-1} (say, A_{n-k-1}) is possible if A_{n-k-1} is obtainable from A_{n-k} by splitting one class in A_{n-k}. The cost of the transition $C(A_{n-k}, A_{n-k-1})$ is defined (in our case) as the sum of the subset heterogeneity measures for the two new classes formed in A_{n-k-1}. An optimal solution is now identified by $\mathcal{F}(A_1)$ for the single entity $A_1 \in \Omega_1$; the optimal hierarchy can again be identified by working backward through the recursion.

Using as an optimization criterion the sum of the $n-2$ subset heterogeneity measures formed while constructing the full partition hierarchy, one could proceed either agglomeratively or divisively, and the same optimal partition hierarchy would be identified. The only major difference in selecting the divisive alternative is that now the partial partition hierarchies provided collaterally from the information yielded by the recursive procedure all start with the n objects contained within one common class and end with a partition into k classes, where k is between 2 and $n-1$. Thus, the divisive approach *begins* with the empirically more useful partitions (having fewer subsets).[22]

3.2.1 Hierarchical Clustering and the Optimal Fitting of Ultrametrics

Ultrametrics:

A concept routinely encountered in discussions of hierarchical clustering is that of an ultrametric, which can be characterized as any nonnegative symmetric dissimilarity matrix for the objects in S, denoted generically by $\mathbf{U} = \{u_{ij}\}$, where $u_{ij} = 0$ if and only if $i = j$ and $u_{ij} \leq \max[u_{ik}, u_{jk}]$ for all $1 \leq i, j, k \leq n$ (this last inequality is equivalent to the statement that for any distinct triple of subscripts, i, j, and k, the largest two proximities among u_{ij}, u_{ik}, and u_{jk} are equal and [therefore] not less than the third). Any ultrametric can be associated with the specific partition hierarchy it induces, having the form $\mathcal{P}_1, \mathcal{P}_2, \ldots, \mathcal{P}_T$, where $\mathcal{P}_1$ and $\mathcal{P}_T$ are now the two trivial partitions that respectively contain all objects in separate classes and all objects in the same class, and $\mathcal{P}_k$ is formed from $\mathcal{P}_{k-1}$ ($2 \leq k \leq T$) by (agglomeratively) uniting certain (and possibly more than two) subsets in $\mathcal{P}_{k-1}$. For those subsets merged in $\mathcal{P}_{k-1}$ to form $\mathcal{P}_k$, all between-subset ultrametric values must be equal and no less than any other ultrametric value associated with an object pair within a class in $\mathcal{P}_{k-1}$. Thus, individual partitions in the hierarchy can be identified by merely increasing a threshold variable starting at zero and observing that $\mathcal{P}_k$ for $1 \leq k \leq T$ is defined by a set of subsets in which all within-subset ultrametric values are less than or equal to some specific threshold value, and all ultrametric values between subsets are strictly greater. Conversely, any partition hierarchy of the form $\mathcal{P}_1, \ldots, \mathcal{P}_T$ can be identified with the equivalence class of all ultrametric matrices that induce it. We note that if only a *single* pair of subsets can be united in $\mathcal{P}_{k-1}$ to form $\mathcal{P}_k$ for $2 \leq k \leq T$, then $T = n$, and we could then revert to the characterization of a full partition hierarchy $\mathcal{P}_1, \ldots, \mathcal{P}_n$ used throughout the previous section.

Given some fixed partition hierarchy, $\mathcal{P}_1, \ldots, \mathcal{P}_T$, there are an infinite number of ultrametric matrices that induce it, but all can be generated by (restricted) monotonic functions of what might be called the basic ultrametric matrix $\mathbf{U}^o = \{u_{ij}^o\}$, whose entries are defined by $u_{ij}^o = \min[k - 1 \mid$ objects O_i and O_j appear within the same class in the partition $\mathcal{P}_k]$. Explicitly, any ultrametric in the equivalence class whose members induce the same fixed hierarchy, $\mathcal{P}_1, \ldots, \mathcal{P}_T$, can be obtained by a strictly increasing monotonic function of the entries in $\mathbf{U}^o$, where the function maps zero to zero. Moreover, because u_{ij}^o for $i \neq j$ can be only one of the integer values from 1 to $T - 1$, each ultrametric in the equivalence class that generates the fixed hierarchy may be defined by one of $T - 1$ distinct values. When these $T - 1$ values are ordered from the smallest to the largest, the $(k - 1)^{st}$ smallest value corresponds to the partition $\mathcal{P}_k$ in the partition hierarchy $\mathcal{P}_1, \ldots, \mathcal{P}_T$, and implicitly to all object pairs that appear together for the first time within a subset in $\mathcal{P}_k$.

To provide an alternative interpretation, the basic ultrametric matrix can also be characterized as defining a collection of linear equality and inequality constraints that any ultrametric in a specific equivalence class must satisfy. Specifically, for each object triple there is (a) a specification of which ultrametric values among the three must be equal, plus two additional inequality constraints so that the third is not greater; (b) an inequality or equality constraint for every pair of ultrametric values based on their order relationship in the basic ultrametric matrix; and (c) an equality constraint of zero for the main diagonal entries in $\mathbf{U}$. In any case, given these fixed equality and inequality constraints, standard L_p regression methods (such as those given in Späth (1991)) could be adapted to generate a best-fitting ultrametric, say, $\mathbf{U}^* = \{u_{ij}^*\}$, to the given proximity matrix $\mathbf{P} = \{p_{ij}\}$. Concretely, we might find $\mathbf{U}^*$ by minimizing

$$\sum_{i<j}(p_{ij} - u_{ij})^2, \quad \sum_{i<j} \mid p_{ij} - u_{ij} \mid, \text{ or possibly } \max_{i<j} \mid p_{ij} - u_{ij} \mid.$$

(As a convenience here and later, it is assumed that $p_{ij} > 0$ for all $i \neq j$, to avoid the technicality of possibly locating best-fitting 'ultrametrics' that could violate the condition that $u_{ij} = 0$ if and only if $i = j$.)

Although we will not pursue this extension in detail, besides defining a collection of linear equality/inequality constraints for identifying best-fitting ultrametrics for a given proximity matrix $\mathbf{P}$, the basic ultrametric matrix $\mathbf{U}^o$ associated with the partition hierarchy $\mathcal{P}_1, \ldots, \mathcal{P}_T$ can also be used to define a collection of linear equality/inequality constraints to generate best-fitting additive-tree metrics to $\mathbf{P}$ that would be consistent with the same partition hierarchy. Briefly, an additive-tree metric can be characterized as any nonnegative symmetric dissimilarity matrix for the objects in S, denoted by $\mathbf{A} = \{a_{ij}\}$, where $a_{ij} = 0$ if and only if $i = j$, and for all $1 \leq i, j, k, l \leq n$, $a_{ij} + a_{kl} \leq \max[a_{ik} + a_{jl}, a_{il} + a_{jk}]$ (or equivalently, among the three sums, $a_{ij} + a_{kl}$, $a_{ik} + a_{jl}$, and $a_{il} + a_{jk}$, the largest two are equal and [therefore] not less than the third). By considering subscript quadruples, i, j, k, and l, and the corresponding sums from $\mathbf{U}^o$ of

$u_{ij}^o + u_{kl}^o$, $u_{ik}^o + u_{jl}^o$, and $u_{il}^o + u_{jk}^o$, the appropriate set of linear equality/inequality constraints is identified for constructing a best-fitting additive-tree metric, say, **A*** (possibly using an L_p regression loss function), that would be consistent with the given partition hierarchy.

Optimal Ultrametrics:

Although an ultrametric matrix may arguably be considered just a convenient device for representing in matrix form the particular partition hierarchy it induces, there has also been a more fundamental usage of the ultrametric notion in the literature as the basic mechanism through which a partition hierarchy might be obtained in the first place. To be explicit, we can develop particular hierarchical clustering methods through direct attempts to find a best-fitting ultrametric for the proximity matrix **P** by the minimization of some criterion, constructed from a function of the discrepancies between the proximities $\{p_{ij}\}$ and the (to be identified) ultrametric matrix $\{u_{ij}\}$ (e.g., an L_p regression alternative might be adopted, as mentioned in connection with fitting an ultrametric to a fixed partition hierarchy). Thus, instead of merely constructing a best-fitting ultrametric subject to a fixed set of linear equality/inequality constraints obtained from a given partition hierarchy, the task becomes one of simultaneously identifying the best set of constraints to impose *and* constructing the best-fitting ultrametric. Most published methods that have attempted in some way to obtain directly a best-fitting ultrametric adopt the least-squares criterion and some auxiliary search strategy for locating an appropriate set of constraints (e.g., see Hartigan (1967); Carroll and Pruzansky (1980); De Soete (1984); Chandon and De Soete (1984); De Soete et al. (1984a), (1984b); Hubert and Arabie (1995b)), although some very notable recent exceptions exist in the use of L_1 and L_∞ norms (e.g., see Farach, Kannan, and Warnow (1995) and Chepoi and Fichet (in press)). For now, we will emphasize the least-squares loss function, but will suggest later how other alternatives might be incorporated.

Following a characterization given by Chandon and De Soete (1984), the optimization task of finding an ultrametric **U*** that will minimize $\sum_{i<j}(p_{ij} - u_{ij})^2$ can be rephrased so as to suggest an application of the GDPP for its solution:

Identify a (full) partition hierarchy, denoted by $\mathcal{P}_1^*, \ldots, \mathcal{P}_n^*$, where

(1) if b_{t-1}^* denotes the average of the proximities *between* the two classes from $\mathcal{P}_{t-1}^*$ that form the new class in $\mathcal{P}_t^*$, then $b_1^* \leq b_2^* \leq \cdots \leq b_{n-1}^*$;

(2) defining the transition cost between two partitions, $T(\mathcal{P}_{t-1}^*, \mathcal{P}_t^*)$, to be the sum of squared deviations of the proximities between the two classes united in $\mathcal{P}_{t-1}^*$ to form the new class in P_t^* from their mean b_{t-1}^*, the partition hierarchy $P_1^*, \ldots, \mathcal{P}_n^*$ minimizes the sum of transition costs given in (3.1) over all possible partition hierarchies that satisfy the condition in (1).

In passing, we note that without loss of generality, the search for an optimal least-squares ultrametric **U*** can be restricted to the use of full partition hierarchies because only weak inequality restrictions are imposed on $b_1^*, \ldots, b_{n-1}^*$. Thus, if an optimal least-squares ultrametric had less than $n-1$ distinct values,

3.2. HIERARCHICAL CLUSTERING

certain of the $b_1^*, \ldots, b_{n-1}^*$ would be tied and would reflect the union of more than two subsets at a particular level in the construction of the optimal partition hierarchy.

It would appear from this rephrasing of how the best-fitting (least-squares) ultrametric might be characterized, that a very direct adoption of the GDPP as applied to the identification of optimal full partition hierarchies would be possible. All that is necessary is to (a) choose a different subset heterogeneity measure in carrying out the recursive process from Ω_1 to Ω_n based on the sum of squared deviations from their mean for those proximities between the two classes united in moving from $\mathcal{P}_{k-1}$ to $\mathcal{P}_k$, and (b) set up a mechanism for imposing the nondecreasing condition on the between-class means themselves through inadmissibility restrictions on certain transitions from one partition to the next. It is this latter addition in (b) that poses some difficulties and also goes directly to the core of the basis for the general validity of a GDPP recursion in the first place.

To be more precise, suppose the new subset heterogeneity measure as described above is adopted, and we are at a point in the recursive process of deciding whether a transition should be admissible in moving from $A_{k-1} \in \Omega_{k-1}$ to $A_k \in \Omega_k$, where A_k is formed from A_{k-1} by uniting two of its subsets. A transition should be inadmissible if the average proximity between the subsets united in A_{k-1} to form A_k is strictly less than the (last) between-subset average that led to A_{k-1}. Unfortunately, and almost inherent in the recursive process, we do not know how A_{k-1} was reached, and therefore, because we do not know the last between-subset average that led to A_{k-1}, inadmissibility cannot be thus defined. Admissibility must instead be characterized using only the partitions A_{k-1} and A_k and their relation to the proximities in $\mathbf{P}$.

The task of defining an admissibility criterion that invariably functions correctly for all proximity matrices, and which uses only the partitions A_{k-1} and A_k, is one that may be insolvable. What can be offered are two less-than-ideal alternatives: (a) an admissibility criterion based only on A_{k-1} and A_k that may sometimes be too lenient and thus fail to ensure that the collection of between-subset aggregate values are nondecreasing for the (purportedly optimal) identified ultrametric, or (b) an admissibility criterion based only on A_{k-1} and A_k that may be too strict, thus yielding the (purportedly optimal) identified ultrametric which could in fact not be the absolute best obtainable. We begin with a detailed discussion of the possibly too lenient admissibility criterion, which is based on observations made by Chandon, Lemaire, and Pouget (1980), and then proceed to the possibly too strict alternative. Later in this section we note the possibility of considering these two criteria in tandem when attempting to identify a least-squares optimal ultrametric for a given proximity matrix $\mathbf{P}$.

Chandon, Lemaire, and Pouget (1980) (see also Chandon and De Soete (1984)), in proposing a branch-and-bound partial enumeration strategy for this same least-squares task of locating a best-fitting ultrametric, offer what initially appears to be a solution for defining admissibility using only the partitions A_{k-1} and A_k and the proximity matrix $\mathbf{P}$. Explicitly, Chandon, Lemaire, and Pouget

(1980) show that a transition is inadmissible if the between-subset average for the two subsets united in A_{k-1} to form A_k is either (a) less than the average of all the proximities *within* the subsets in A_{k-1} (or more strongly [although Chandon, Lemaire, and Pouget (1980) do not give this result explicitly], inadmissibility could be generalized to the between-subset average being less than the average proximity within any subset in A_{k-1}), or (b) greater than the average of all proximities *between* the subsets in A_k. Unfortunately, these two conditions characterizing possible inadmissibility are not always sufficient to ensure a nondecreasing sequence of the between-subset averages, and a counterexample can be provided from the proximity matrix among the first six digits chosen from Table 1.1. Using the subset heterogeneity measure, defined as the sum of squared deviations from their mean for the proximities between the subsets united in moving from $\mathcal{P}_{k-1}$ to $\mathcal{P}_k$, and imposing the two inadmissibility criteria given above, the optimal partition hierarchy given below would have a value for the least-squares loss criterion of .10004 (the mean proximity between the two subsets united at each level to form the successive partitions is also provided). There is an obvious violation of the nondecreasing pattern needed for the mean proximities between subsets because the level 4 value of .371 is less than the level 3 value of .465. However, both the partitions at levels 3 and 4 are respectively admissible transformations from levels 2 and 3 using the two inadmissibility criteria discussed by Chandon, Lemaire, and Pouget (1980). Explicitly, for the level 3 partition, the between-subset value of .465 is greater than the average proximity of .059 for those objects already united at level 2 and less than the average proximity of .534 between objects not already united at level 3; similarly, for the level 4 partition, the between-subset value of .371 is greater than the average proximity of .330 for those objects already united at level 3, and less than the average proximity of .589 between objects not already united at level 4.

A counterexample to the use of the Chandon, Lemaire, and Pouget admissibility criteria, obtained using DPHI1U:

Level	Partition	Mean Proximity Between United Subsets
1	(all digits separate)	—
2	{{2,4},{0},{1},{3},{5}}	.059
3	{{1,2,4},{0},{3},{5}}	.465
4	{{1,2,3,4},{0},{5}}	.371
5	{{1,2,3,4,5},{0}}	.524
6	(all digits together)	.640

The basic difficulty with the Chandon, Lemaire, and Pouget (1980) inadmissibility criteria is that they may not be restrictive enough to eliminate (nontrivial) order inversions in the sequence of between-subset averages when progressing from $\mathcal{P}_1$ to $\mathcal{P}_n$. Here, a *nontrivial* order inversion is one in which the average between-subset proximity for the formation of a new subset at some level, say, k, is strictly greater than the average between-subset proximity involving this new subset at level k and some other subset, with the two united to form another

3.2. HIERARCHICAL CLUSTERING

new subset at level k', where $k < k'$. The above example provides an illustration of a nontrivial order inversion in the two new subsets formed at levels 3 and 4. A trivial order inversion, on the other hand, can effectively be ignored because it can always be eliminated by just rearranging in some way the order of new subset formation at different levels of the partition hierarchy. In the following example, for example, if the formation of the two disjoint subsets at levels 4 and 5 were reversed so {2,4,8} was formed at level 4 and {3,6,9} was formed at level 5, the order inversion would be trivial and easily eliminated by using the order given for the formation of these two subsets.

The two Chandon, Lemaire, and Pouget (1980) criteria may turn out to be sufficient for a particular data set to eliminate all nontrivial order inversions. For example, merely using these criteria on the complete set of ten digits produced the following optimal partition hierarchy where the between-subset averages are nondecreasing (no single object sets are listed at each level below). The least-squares loss function for this hierarchy has a value of .4698, and a corresponding 'variance-accounted-for' of .4941 (using the usual 'variance-accounted-for' formula of

$$1 - \left[\sum_{i<j}(p_{ij} - u^*_{ij})^2 / \sum_{i<j}(p_{ij} - \bar{p})^2 \right],$$

where $\bar{p}$ is the mean proximity of all off-diagonal entries in $\mathbf{P}$, and u^*_{ij} is the least-squares optimal ultrametric value for objects O_i and O_j). Interpretively, and consistent with the partition hierarchies used as illustrations earlier, the structural characteristics of the digits are apparent up through the partition present at level 6, with numerical magnitude being depicted more in the partitions constructed beyond level 6.

An optimal partition hierarchy based on the least-squares estimation of a best-fitting ultrametric, obtained using DPHI1U:

Level	Partition	Mean Proximity Between United Subsets
1	(all digits separate)	—
2	{2,4}	.059
3	{2,4},{3,9}	.263
4	{2,4},{3,6,9}	.298
5	{2,4,8},{3,6,9}	.307
6	{2,4,8},{3,6,9},{5,7}	.400
7	{2,4,8},{3,5,6,7,9}	.481
8	{2,3,4,5,6,7,8,9}	.553
9	{1,2,3,4,5,6,7,8,9}	.584
10	(all digits together)	.730

There is an inadmissibility restriction that could be imposed (in addition to, or instead of, those suggested by Chandon, Lemaire, and Pouget (1980)) that would be sufficient to guarantee a nondecreasing set of between-subset averages, but it could possibly be too restrictive. Explicitly, in considering a transition from $\mathcal{P}_{k-1}$ to $\mathcal{P}_k$, the average between-subset proximity for the two subsets united in $\mathcal{P}_{k-1}$ could be required to be no greater than the average proximity between the newly formed subset in $\mathcal{P}_k$ and *any other* subset in $\mathcal{P}_k$. As noted above, it is possible that this criterion may be too restrictive for some proximity matrices in the sense that an optimal least-squares ultrametric may not satisfy it (i.e., for some subset in $\mathcal{P}_k$, the average proximity to the newly formed subset in $\mathcal{P}_k$ could be less than the average between-subset average that led to the newly formed subset in $\mathcal{P}_k$; however, in the optimal least-squares ultrametric, that specific subset is not directly united with the newly formed subset in $\mathcal{P}_k$ but is first merged with other subset(s) in $\mathcal{P}_k$ before an eventual merging with the new subset that defined $\mathcal{P}_k$).

Both these two options for defining inadmissibility: the Chandon, Lemaire, and Pouget (1980) criteria, which may be too lenient for some proximity matrices, and the one just proposed, which possibly could be too restrictive, can be used complementarily. As a suggested analysis strategy, if no nontrivial order inversions appear in the partition hierarchy with the Chandon, Lemaire, and Pouget (1980) option, a least-squares best-fitting ultrametric has then been identified. If, on the other hand, nontrivial order inversions do appear, then the stricter inadmissibility criterion could be used, and a best-fitting least-squares hierarchy finally identified (hopefully) — but with the caveat that in some (presumably rare) instances an even better one could be constructed.[23]

The GDPP strategy for locating best-fitting ultrametrics that are least-squares optimal can be extended directly to the use of loss functions based on L_p norms other than least-squares. For example, if the optimal ultrametric $\mathbf{U}^*$ is to minimize the L_1 loss function, $\sum_{i<j} \mid p_{ij} - u^*_{ij} \mid$, rather than the L_2 least-squares criterion of $\sum_{i<j}(p_{ij} - u^*_{ij})^2$, the use of average between-subset proximities would be replaced by the *median* between-subset proximity, and the transition cost between two partitions, $T(\mathcal{P}^*_{k-1}, \mathcal{P}^*_k)$, would be defined by the sum of absolute deviations from their median for the proximities between the two classes united in $\mathcal{P}^*_{k-1}$ to form the new class in $\mathcal{P}^*_k$. Or, if the average between-subset proximity is replaced by the average of the minimum and maximum between-subset proximities, and the transition cost $T(\mathcal{P}^*_{k-1}, \mathcal{P}^*_k)$ specified as the maximum deviation from this value for those proximities between the two classes in $\mathcal{P}^*_{k-1}$ united to form $\mathcal{P}^*_k$, the loss function would be one of minimizing the sum of the maximum of such deviations over the $n-2$ new subsets formed in constructing an optimal partition hierarchy $\mathcal{P}^*_1, \ldots, \mathcal{P}^*_n$. We will not explicitly provide numerical examples for such generalizations here (although these L_p extensions are included as options in the heuristic modifications [and associated computer programs] discussed in Chapter 5), and instead refer the reader to Hubert, Arabie, and Meulman (1997c) for a further development and illustration of these ideas in the hierarchical clustering context.

Optimal ultrametrics based on dissimilarity order. Besides extending the GDPP strategy to identify best-fitting L_p-norm ultrametrics, a variety of different loss functions could be considered that would depend *only* on the ordering of proximities in $\{p_{ij}\}$ and the ordering of the ultrametric values in the (to-be-identified) ultrametric matrix $\{u_{ij}\}$. We mention two possibilities here that

3.2. HIERARCHICAL CLUSTERING

involve minimizing the number of order inconsistencies, of particular kinds, between $\{p_{ij}\}$ and the to-be-identified base ultrametric $\{u_{ij}^o\}$ (a base ultrametric can be used without loss of generality because only the order properties are being considered in the entries for the ultrametric matrix). First, we define an object triple inconsistency (OTI) for the subscripts i, j, and k whenever the expression $\text{sign}((u_{ij}^o - u_{ik}^o)(p_{ij} - p_{ik}))$ is strictly negative, where $\text{sign}(x) = +1, 0, -1$ when $x > 0, = 0$, or < 0, respectively. Obviously, an OTI occurs for a particular subscript triple, i, j, and k, whenever the ordering of the two dissimilarities, p_{ij} and p_{ik}, is opposite to what a good-fitting base ultrametric should suggest, i.e., when $p_{ij} > p_{ik}$ but $u_{ij}^o < u_{ik}^o$, or $p_{ij} < p_{ik}$ but $u_{ij}^o > u_{ik}^o$. This OTI index, when summed over object triples, bears a clear similarity to what is counted in the count b heterogeneity measure for a single partition (as defined and used in Section 3.1)—inconsistencies in both cases are counted for two pairs of indices only when a common object is included.

To use OTI to define the transition cost $T(\mathcal{P}_{k-1}^*, \mathcal{P}_k^*)$, we consider the two classes united in $\mathcal{P}_{k-1}^*$ (to form the new class in $\mathcal{P}_k^*$), and pick two indices, say, i' and j', where i' and j' belong to (or span) the separate classes and k' is an index outside *both* the two classes. If $p_{i'j'} > p_{i'k'}$, then an OTI is present for the particular subscript triple i', j', and k', and counting the number of such inconsistencies (OTIs) over all such triples thus definable, the transition cost $T(\mathcal{P}_{k-1}^*, \mathcal{P}_k^*)$ is obtained. This cost is thus generated only by those "new" OTIs that would be produced between the two classes in $\mathcal{P}_{k-1}^*$ and their union, but not from the OTIs that may be definable separately within these two classes, and which presumably are counted earlier in the hierarchy.

Second, we define the number of object quadruple inconsistencies (OQIs) by the number of times $\text{sign}((u_{ij}^o - u_{kl}^o)(p_{ij} - p_{kl}))$ is strictly negative over all $1 \leq i, j, k, l \leq n$. Just as the OTIs were analogous to count b for single partitions, the OQIs can be viewed similar to the count a heterogeneity measure for single partitions (again, as defined and used in Section 3.1) — counts will be made of the number of instances in which proximities within subsets are strictly greater than those between the subset and its complement (and without the mandatory common object required in identifying an OTI). To use OQIs to define transition cost, and extending the discussion above for OTIs for the two classes to be united in $\mathcal{P}_{k-1}^*$, the two indices i' and j' span the separate classes, and for any two other indices k' and l', either k' or l' must be in the union of the separate classes and with the other index outside. If $p_{ij} > p_{kl}$, then a quadruple inconsistency is present for the particular subscript quadruple i', j', k', and l', and counting the number of such inconsistencies over all such quadruples thus defined, the transition cost $T(\mathcal{P}_{k-1}^*, \mathcal{P}_k^*)$ is obtained.

To illustrate the use of the OTI and OQI indices, optimal full partition hierarchies and partial partition hierarchies up to four subsets are given below. As is apparent, for both full partition hierarchies we have interpretations that require some rather annoying mixtures of digit magnitude and digit structural characteristics. Both partial partition hierarchies, however, are very clearly interpretable by just the esthetic, structural characteristics of the digits. We may have an empirical exemplar here that, forcing optimality for identifying a full partition hierarchy, may be counterproductive vis-à-vis an ultimate final substantive interpretation. The partial partition hierarchies are much easier to explain, and are still optimal, although to a limited level in the hierarchy.[24]

An optimal full partition hierarchy and a partial partition hierarchy up to four subsets based on minimizing the number of triple inconsistencies (OTIs) between a base ultrametric and the given proximity matrix, obtained using HPHI1U, is given in the following table:

Level	Partition	Cumulative OTIs
1	(all digits separate)	—
2	{1,2}	1
3	{0,1,2}	6
4	{0,1,2},{3,9}	6
5	{0,1,2},{3,6,9}	6
6	{0,1,2},{3,6,9},{4,8}	7
7	{0,1,2},{3,4,6,8,9}	28
8	{0,1,2},{3,4,6,8,9},{5,7}	29
9	{0,1,2},{3,4,5,6,7,8,9}	42
10	(all digits together)	42
1	(all digits separate)	—
2	{0,1}	2
3	{0,1},{2,4}	2
4	{0,1},{2,4},{3,9}	2
5	{0,1},{2,4},{3,6,9}	2
6	{0,1},{2,4},{3,6,9},{5,7}	3
7	{0,1},{2,4,8},{3,6,9},{5,7}	6

An optimal full partition hierarchy and a partial partition hierarchy up to four subsets based on minimizing the number of quadruple inconsistencies (OQIs) between a base ultrametric and the given proximity matrix, obtained using HPHI1U, is given in the following table:

Level	Partition	Cumulative OQIs
1	(all digits separate)	—
2	{0,1}	2
3	{0,1,2}	8
4	{0,1,2,4}	32
5	{0,1,2,3,4}	58
6	{0,1,2,3,4},{5,7}	59
7	{0,1,2,3,4},{5,6,7}	69
8	{0,1,2,3,4},{5,6,7},{8,9}	74
9	{0,1,2,3,4,},{5,6,7,8,9}	113
10	(all digits together)	113
1	(all digits separate)	—
2	{0,1}	2
3	{0,1},{2,4}	2

4	{0,1},{2,4},{3,9}	2
5	{0,1},{2,4},{3,6,9}	2
6	{0,1},{2,4},{3,6,9},{5,7}	3
7	{0,1},{2,4,8},{3,6,9},{5,7}	6

3.2.2 Constrained Hierarchical Clustering

In the application of the GDPP to the task of hierarchical clustering (in Section 3.2), the n sets $\Omega_1, \ldots, \Omega_n$ on which the recursive process was based contained for Ω_k all the possible partitions of the n objects in S into $n - k + 1$ classes. Because of the large number of such partitions at each level for even moderate values of n and the need to store and access information for each partition during the recursive process, the size of the data sets that can be analyzed has an effective limit of n in the lower teens. Analogous to the discussion in Section 3.1.1 on admissibility restrictions in constructing single partitions, if one could limit the type of partition to be considered at each level of the hierarchy and redefine the sets $\Omega_1, \ldots, \Omega_n$ accordingly, it may be possible to deal effectively with somewhat larger object set sizes.

Analogous also to the admissibility conditions discussed for partitions in Section 3.1.1, one constraint that might be imposed on each partition in Ω_k is for the constituent subsets to contain objects consecutive in some given ordering (which could be taken as $O_1 \prec O_2 \prec \cdots \prec O_n$ without loss of any generality). Thus, Ω_k would be redefined to contain those partitions that include $n - k + 1$ classes, where each class is defined by a segment in the given object ordering.[25]

A numerical illustration. To give a brief example of constrained hierarchical clustering, and one that can also be compared to an unconstrained result given earlier in Section 3.2, an optimal partition hierarchy using the diameter criterion (the subset heterogeneity measure in (iv)) is given below. Here, an order constraint is imposed that subsets in a partition must contain digits consecutive in magnitude.

A (restricted) optimal partition hierarchy based on the sum of subset diameters, obtained using DPHI1R:

Level	Partition	Index of New Subset	Cumulative Sum
1	(all digits separate)	—	—
2	{1,2}	.284	.284
3	{1,2,3}	.354	.638
4	{1,2,3},{8,9}	.392	1.030
5	{1,2,3},{5,6},{8,9}	.396	1.426
6	{1,2,3},{4,5,6},{8,9}	.409	1.835
7	{1,2,3},{4,5,6},{7,8,9}	.459	2.294
8	{0,1,2,3},{4,5,6},{7,8,9}	.709	3.003
9	{0,1,2,3},{4,5,6,7,8,9}	.742	3.745
10	(all digits together)	.909	4.654

The unconstrained partition hierarchy given earlier has an (obviously) better cumulative sum of 4.119 compared to the current value of 4.654 when the subsets are constrained.[26]

Endnotes

[7]The use of a proximity matrix where the only nonmissing entries are between two disjoint sets will reappear in Chapter 4 (Section 4.1.3). The optimization task introduced there is discussed under the label of unfolding, which involves (in its one-dimensional form) the joint sequencing of two disjoint object sets along a common continuum.

[8]The first published version of this particular DP strategy for minimizing the sum of heterogeneity measures over the classes of a partition was by Jensen (1969), but only for the specific heterogeneity measure (as we note below) that defines the sum of squared error criterion (or what is commonly called the k-means criterion; see MacQueen (1967)).

[9]The notation of $\max(S_m)$ and $\min(S_m)$ is used here only as a convenience in referring to the heterogeneity for some fixed set S_m. It is not to be interpreted as an index that we are trying to optimize by the choice of S_m.

[10]Given that **P** is symmetric, an obvious alternative choice for defining these various sum($\cdot$) expressions would have counted each distinct object pair only once rather than twice. Nothing done in what follows involving optimization depends on whether distinct object pairs are counted once or twice. Either the measures being minimized would be identical when individual subset heterogeneity indices are given by averages, or one would be twice the other when raw sums are used.

[11]Note that the calculation of count$a$ requires considering four objects at a time, whereas count$b$ requires only three. Also, every instance that contributes an increment to count$b$ will contribute to count$a$, but not conversely.

[12]The program DPCL1U (where the suffix '1U' denotes '1-mode unrestricted') includes all the heterogeneity indices described and the two options for the optimization criterion of minimizing either their sum or their maximum over the M subsets of a partition, and was used to carry out the illustrative analyses reported. DPCL1U is written (potentially) to allow the analyses of object set sizes up to 30 using allocatable arrays, but for n above, say, 20, a very large amount of RAM is required. Object set sizes that are less than or equal to 20 are more in the range of commonly available storage configurations. Using DPCL1U to partition an object set of size, say, 27, would require 4GB of RAM and would be at the addressable limit for a 32 bit operating system.

[13]There is some terminological confusion in the literature about the use of "Robinson" versus "anti-Robinson." When Robinson (1951) published his classic paper on chronologically ordering archaeological deposits by identifying a row and column ordering of a proximity matrix to produce (approximately) a characteristic pattern in the reordered matrix, he explicitly dealt only with *similarity* matrices. Thus, the keying of the proximities considered by Robinson was such that the larger values referred to the more similar objects (in

contrast to a keying as a *dissimilarity* in which large values refer to the less similar objects). The pattern Robinson (1951) sought to identify in a reordered similarity matrix was one in which the entries within each row and column never *in*creased moving away from a main diagonal entry in any direction. In work published in the 1960's and early 1970's, and in honor of Robinson's seminal contributions, the name "Robinson" was attached to the pattern expected when reordering a *similarity* matrix, and *only for a similarity matrix*. There is no better source than the now classic volume edited by Hodson, Kendall, and Tăutu (1971) and devoted to *Mathematics in the Archaeological and Historical Sciences*. The many papers therein consistently define and reserve the term "Robinson" for a pattern expected in a similarity matrix, e.g., see Gelfand (1971), Kendall (1971b), Sibson (1971), and Landau and de la Vega (1971). Although the need to distinguish between Robinson and anti-Robinson forms for proximity matrices respectively keyed as similarities and dissimilarities was pointed out explicitly in the 1970's (e.g., see Hubert (1976), p. 33), some subsequent authors have adopted the terms "Robinson" or "Robinsonian" to refer to patterns in dissimilarity matrices without distinguishing the reversed pattern thus implied in the matrix entries; e.g., see Diday (1986), Batagelj, Korenjak-Černe, and Klavžar (1994), and Critchley (1994), among others. Authors more respectful of the discordant patterns and associated types of proximity data have maintained (some might say, stubbornly) the distinction, e.g., Hubert and Arabie (1994) in their discussion of fitting (analyzing) proximity matrices through sums of matrices each having the (anti-)Robinson form, or Hubert, Arabie, and Meulman (1998) in developing graph-theoretic representations for proximity matrices through strongly-anti-Robinson or circular strongly-anti-Robinson matrices. In any case, because all the proximity matrices considered throughout this monograph are keyed as dissimilarity matrices, we will consistently use the term "anti-Robinson" whenever appropriate to refer to the pattern expected in a proximity matrix. In Section 4.1 the issue of representing an (anti-)Robinson matrix graphically will be revisited (in an endnote) and some comments made about the difficulties that may be encountered if the anti-Robinson condition is not strengthened (to what is called strongly-anti-Robinson).

[14] An implementation of order-constrained clustering is available in a program called DPCL1R (where the suffix '1R' denotes '1-mode restricted'), which, analogous to DPCL1U, includes the use of all the same subset heterogeneity options and the two optimization criteria (i.e., either minimizing the sum or the maximum). The specific object order that defines the admissible clustering can either be the identity order or be specified by the user at run time. Again, the use of allocatable arrays will determine whether the memory capacity of the system is sufficient to solve the problem of the size requested, but here the tasks can be rather large for the typical amount of RAM, e.g., object set sizes up to, say, 500.

[15] Exactly this approach is carried out in the program DPCL2U (where the suffix '2U' now denotes '2-mode unrestricted'); this program is a direct generalization of DPCL1U in carrying out an unconstrained optimization approach to the partitioning of $S_A \cup S_B$.

[16] As an aside, if for the LA task discussed in the introductory section, Section 2.1, the $n \times n$ cost matrix $\{c_{ij}\}$ is treated as a proximity matrix between its n rows and n columns, the LA task in a minimization form (for either the sum of costs or the maximum cost over the $n!$ assignments) is solved as a special case of the partitioning task if the maximum number of subsets, n, is used and the subset heterogeneity measure is defined, among other possibilities, by the within-subset average.

[17] There are no built-in size limits in DPCL2R (where the suffix '2R' now denotes '2-mode restricted'), allowing the use of dual row and column object orders. The program is otherwise the analogue of DPCL1R; however, given the need for large allocatable arrays and the size of most current systems' RAM, the effective limit may be about 100 for the sum of n_A and n_B.

[18] Given the notion that a common storage configuration includes somewhere between 64 and 128 MB of RAM, object sets of size 12 and under are reasonable to approach with the GDPP. There are, however, a huge number of partitions definable for object sets (just) larger than 12. For example, based on the Stirling numbers of the second kind, there are about 4 million partitions of an object set of size 12, but about 200 million for an object set of size 14. Thus, for an object set of size 14, 2GB of RAM would be necessary just for storing the information for each partition in the course of carrying out the recursive process.

[19] The program DPHI1U, used to obtain optimal full partition hierarchies, routinely provides all optimal partial hierarchies for $k = 2, \ldots, n-1$.

[20] Thus, we have an empirical demonstration that partial partition hierarchies identifiable within a globally optimal full partition hierarchy may not necessarily be optimal partial partition hierarchies themselves. Or stated differently, optimal partial partition hierarchies are not necessarily nested in the sense that an optimal partial hierarchy that terminates at level $t-1$ is not necessarily included as part of an optimal partial hierarchy that terminates at level t.

[21] Although we will not pursue the point in any detail, we note that *if* we were considering *only* partitions with four classes, such as the level 7 partition identified in the optimal hierarchy given above, which has a maximum diameter over all four subsets of .421, there is another partition with four classes that also has a maximum diameter of .421, but which is completely consistent with number magnitude, i.e., $\{\{0,1\},\{2,3,4\},\{5,6,7\},\{8,9\}\}$.

[22] The divisive analogue of the agglomerative strategy implemented in the program DPHI1U is available in a separate program, called DPDI1U, with all the same options for the choice of the subset heterogeneity measure.

[23] Both inadmissibility options are included in the programs DPHI1U and DPDI1U.

[24] The heuristically oriented programs HPHI1U and HPHI2U, as discussed further in Chapter 5 (where the leading letter 'H' denotes 'heuristic'), include both options of minimizing the number of OTIs or OQIs in complete or partial partition hierarchy construction. As a special case for small object sets, such as the one illustrated here, optimal solutions are possible with these two programs.

3.2. HIERARCHICAL CLUSTERING

[25]Such an option is available in the program DPHI1R (where the suffix '1R' again denotes '1-mode restricted'); this approach is a direct extension of DPHI1U with all the latter's options, but also requires a specific ordering of the objects to be used in characterizing those partitions considered admissible when carrying out the recursive process from Ω_1 to Ω_n. Although there is a formal limit of 30 built into DPHI1R for the object set size, most current RAM configurations (as of, say, 1999) would generally allow n to be about 20.

[26]Special diagnostics would have to be developed to decide whether such an increase in loss should be considered intolerably large.

Chapter 4

Object Sequencing and Seriation

"Things are ordered in a wonderful manner."
Joseph Conrad
Under Western Eyes (1911)

"...in consideration of the imperfection inherent in the order of the world..."
Heinrich von Kleist
The Marquise of O (1808)
Translation by Nigel Reeves

" '...so Princeton,' Dawn said, 'so *unerring*. He works so hard to be one-dimensional...' "
Philip Roth
American Pastoral (1997)

The three major sections of this chapter discuss various aspects of the task of optimally sequencing (or seriating) a set S of n objects along a continuum based on whatever proximity data may be available between the object pairs. The ultimate purpose of any sequencing technique, as is generally true for all CDA methods, is to use whatever combinatorial structure is identified (which in the present context will be object orderings) to help explain the relationships that may be present among the objects, reflected by the proximity data. Section 4.1 is concerned with the (unconstrained) sequencing of a *single* object set, although this single set may itself be the union of two disjoint sets. In this latter case, the task of object placement is one of obtaining a joint sequencing of the two sets (i.e., we have a data analysis task usually discussed under the subject of unidimensional unfolding; e.g., see Heiser (1981)). In general, proximity data may be in the form of a one-mode matrix that may either be symmetric (which would be consistent with the discussion in the clustering framework of Chapter 3) or nonsymmetric (and more specifically, skew-symmetric), or possibly in the form of a two-mode matrix if the set S is formed by the union of two other sets. Section 4.2 discusses only two-mode proximity data, within a context where if the single object set S is formed from the union of the sets S_A and S_B, the ordering of the objects *within* S_A and/or *within* S_B is constrained in their joint placement along a continuum. In effect, we are combining (or in a slightly different interpretive sense, we are comparing) the two distinct object sets, subject to some type of constraint on the orderings of the objects within S_A and S_B. The final Section 4.3 provides an optimization problem constituting a mixture of an unconstrained sequencing task, as discussed in Section 4.1, and the clustering problem of Section 3.1. Specifically, the goal will be to construct optimal partitions of an object set S in which the classes of the partition must themselves be ordered along a continuum. Obviously, when the number of classes in the partition is set equal to the number of objects in S, this latter optimization task reduces to the sequencing of a single object set as described in Section 4.1.

In the topics outlined briefly above that are to be pursued in the sections to follow, three types of proximity data will organize the presentation. One type comes in the usual form of a one-mode $n \times n$ nonnegative dissimilarity matrix for the n objects in S denoted earlier by $\mathbf{P} = \{p_{ij}\}$, where $p_{ij} = p_{ji} \geq 0$ and $p_{ii} = 0$ for $1 \leq i, j \leq n$. A second type is an $n_A \times n_B$ two-mode nonnegative dissimilarity matrix $\mathbf{Q}$, where proximity is defined only between the objects from two distinct sets S_A and S_B containing, respectively, n_A and n_B objects and forming the rows and columns of $\mathbf{Q}$. (For convenience, the objects in S_A are again denoted as $\{r_1, \ldots, r_{n_A}\}$ and those in S_B as $\{c_1, \ldots, c_{n_B}\}$, where the letters 'r' and 'c' signify 'row' and 'column', respectively.) In the two-mode context, it will typically be convenient to construct a single mode from the objects in both S_A and S_B (i.e., $S \equiv S_A \cup S_B$), and use $\mathbf{Q}$ to generate a nonnegative symmetric proximity matrix on S that contains missing proximities for object pairs contained within S_A or within S_B. To refer to this latter matrix explicitly, we introduce the notation $\mathbf{P}^{AB} = \{p_{ij}^{AB}\}$, where $\mathbf{P}^{AB}$ has the block form

$$\begin{bmatrix} * & \mathbf{Q} \\ \mathbf{Q}' & * \end{bmatrix}$$

and the asterisk * denotes those proximities that are missing.

Finally, if the proximity information originally given between the pairs of objects in S is nonsymmetric and in the form of an $n \times n$ nonnegative matrix, say, $\mathbf{D} = \{d_{ij}\}$, this latter matrix will first be decomposed into the sum of its symmetric and skew-symmetric components,

$$\mathbf{D} = [(\mathbf{D} + \mathbf{D}')/2] + [(\mathbf{D} - \mathbf{D}')/2],$$

and each of these components will (always) be addressed separately. The matrix $(\mathbf{D} + \mathbf{D}')/2$ can be treated merely as a nonnegative symmetric dissimilarity matrix (i.e., in the notation $\mathbf{P}$), and methods appropriate for such a proximity measure can be applied. The second component of the form $(\mathbf{D} - \mathbf{D}')/2$ is an $n \times n$ matrix denoted by $\mathbf{P}^{SS} = \{p_{ij}^{SS}\}$, where $p_{ij}^{SS} = (d_{ij} - d_{ji})/2$ for $1 \leq i, j \leq n$, and the superscript 'SS' signifies 'skew-symmetric', i.e., $p_{ij}^{SS} = -p_{ji}^{SS}$ for all $1 \leq i,j \leq n$. Thus, there is an explicit directionality to the (dominance) relationship specified between any two objects depending on the order of the subscripts, i.e., p_{ji}^{SS} and p_{ij}^{SS} are equal in absolute magnitude but differ in sign. It may also be worth noting here that any skew-symmetric matrix $\{p_{ij}^{SS}\}$ can itself be interpreted as containing two separate sources of information: one is the directionality of the dominance relationship given by $\{\text{sign}(p_{ij}^{SS})\}$, where (as defined earlier) $\text{sign}(y) = 1$ if $y > 0$; $= 0$ if $y = 0$; and $= -1$ if $y < 0$; the second is in the absolute magnitude of dominance given by $\{|\,p_{ij}^{SS}\,|\}$. This latter matrix is symmetric and can be viewed as a dissimilarity matrix and analyzed as such. In fact, the first numerical examples below on object sequencing for a one-mode symmetric proximity matrix will use the paired-comparisons data from Table 1.2 in exactly this manner for constructing optimal orderings of the offenses according to their perceived seriousness.

4.1 Optimal Sequencing of a Single Object Set

The search for an optimal sequencing of a single object set S (irrespective of the type of proximity measure available and whether S itself is the union of two other sets) can be operationalized by constructing a best reordering for the rows and simultaneously the columns of an $n \times n$ proximity matrix. The row/column reordering to be identified will optimize, over all possible row/column reorderings, some specified measure of patterning for the entries of the reordered matrix. For convenience, the particular measure of pattern will usually be defined so as to be maximized (with a few exceptions, as in Section 4.1.4), and thus, the general form of the GDPP to be applied is the recursion given in (2.3). (Again, much as in the hierarchical clustering framework, the application of the max/min type of recursion in (2.5) has limited utility. When constructing an optimal ordering based on the type of measure of matrix patterning we will consider, the various positions in an ordering will have differential possible contributions to the measure of matrix pattern. As a consequence, a max/min criterion would overemphasize the placement of a very few middle objects in the ordering and effectively ignore how well the entire early and late portions of a sequence were constructed. We note, however, that not all possible measures of matrix patterning would have this property of a differential possible contribution depending on location in an ordering. Specifically, there is a particular application of a max/min (or min/max) recursion for obtaining a best sequencing through the construction of optimal paths that will be discussed later in Section 4.1.4.)

A variety of specific measures of patterning will be introduced in the next three sections within the context of the particular type of proximity measure most appropriate (i.e., symmetric ($\mathbf{P}$), skew-symmetric ($\mathbf{P}^{SS}$), or defined between two sets S_A and S_B ($\mathbf{P}^{AB}$)). However, in all three cases the same specialization of the GDPP in (2.3) will be implemented. (We should again note that a variation is introduced in Section 4.1.4 in a discussion of object sequencing based on the construction of optimal paths.)

A collection of sets $\Omega_1, \ldots, \Omega_n$ is defined (so, $K \equiv n$), where Ω_k includes all the subsets that have k members from the integer set $\{1, 2, \ldots, n\}$. The value $\mathcal{F}(A_k)$ is the optimal contribution to the total measure of matrix patterning for the objects in A_k when they occupy the first k positions in the (re)ordering. A transformation is now possible between $A_{k-1} \in \Omega_{k-1}$ and $A_k \in \Omega_k$ if $A_{k-1} \subset A_k$ (i.e., A_{k-1} and A_k differ by one integer). The contribution to the total measure of patterning generated by placing the single integer in $A_{k-1} - A_k$ at the kth order position is $M(A_{k-1}, A_k)$. As always, the validity of the recursive process will require the incremental merit index $M(A_{k-1}, A_k)$ to depend only on the unordered sets A_{k-1} and A_k and the complement $S - A_k$, and specifically *not* on how A_{k-1} may have been reached beginning with Ω_1. Assuming $\mathcal{F}(A_1)$ for all $A_1 \in \Omega_1$ are available, the recursive process can be carried out from Ω_1 to Ω_n, with $\mathcal{F}(A_n)$ for the single set $A_n = \{1, 2, \ldots, n\} \in \Omega_n$ defining the optimal value for the specified measure of matrix patterning. The optimal row/column reordering is constructed, as always, by working backward through the recursion.[27]

4.1.1 Symmetric One-Mode Proximity Matrices

When the original data come in the form of a nonnegative symmetric proximity matrix **P** with no missing entries, there are (as might be expected) many different indices of patterning that could be optimized in a row/column reordered proximity matrix through the specialization of the GDPP described above. We will emphasize two general classes of such measures below.[28]

Row and/or column gradient measures. One ubiquitous concept encountered in the literature on matrix reordering is that of a symmetric proximity matrix having an anti-Robinson form (this same structure was noted briefly in the clustering context when an optimal unconstrained clustering might also be optimal when order-constrained). Specifically, suppose $\rho(\cdot)$ is some permutation of the first n integers that reorders both the rows and columns of **P** (i.e., $\mathbf{P}_\rho \equiv \{p_{\rho(i)\rho(j)}\}$). The reordered matrix $\mathbf{P}_\rho$ is said to have an anti-Robinson form if the entries within the rows and within the columns of $\mathbf{P}_\rho$ moving away from the main diagonal in any direction never decrease; or formally, two gradient conditions must be satisfied:

within rows: $p_{\rho(i)\rho(k)} \leq p_{\rho(i)\rho(j)}$ for $1 \leq i < k < j \leq n$;
within columns: $p_{\rho(k)\rho(j)} \leq p_{\rho(i)\rho(j)}$ for $1 \leq i < k < j \leq n$.

We might note that whenever **P** is an ultrametric, or if **P** has an exact Euclidean representation in a single dimension (i.e., $\mathbf{P} = \{|x_j - x_i|\}$ for some collection of coordinate values, $x_1, x_2, \ldots, x_n$), then **P** can be row/column reordered to display a perfect anti-Robinson pattern. Thus, the notion of an anti-Robinson form can be interpreted as generalizing either a perfect discrete classificatory structure induced by a partition hierarchy (through an ultrametric) or as the pattern expected in **P** if there exists an exact unidimensional

4.1. OPTIMAL SEQUENCING OF A SINGLE OBJECT SET

Euclidean representation for the objects in S. In any case, if a matrix can be row/column reordered to display an anti-Robinson form, then the objects are orderable along a continuum so that the degree of separation between objects in the ordering is reflected perfectly by the dissimilarity information in $\mathbf{P}$, i.e., for the object ordering, $O_{\rho(i)} \prec O_{\rho(k)} \prec O_{\rho(j)}$ (for $i < k < j$), $p_{\rho(i)\rho(k)} \leq p_{\rho(i)\rho(j)}$ and $p_{\rho(k)\rho(j)} \leq p_{\rho(i)\rho(j)}$ (or equivalently, because the most extreme separation in the ordering of the three distinct objects is between $O_{\rho(i)}$ and $O_{\rho(j)}$, we have $p_{\rho(i)\rho(j)} \geq \max[p_{\rho(i)\rho(k)}, p_{\rho(k)\rho(j)}])$.[29]

A natural (merit) measure for how well the particular reordered proximity matrix $\mathbf{P}_\rho$ satisfies these two gradient conditions would rely on an aggregate index of the violations/nonviolations over all distinct object triples, as given by the expression

$$\sum_{i<k<j} f(p_{\rho(i)\rho(k)}, p_{\rho(i)\rho(j)}) + \sum_{i<k<j} f(p_{\rho(k)\rho(j)}, p_{\rho(i)\rho(j)}), \qquad (4.1)$$

where $f(\cdot, \cdot)$ is some function indicating how a violation/nonviolation of a particular gradient condition for an object triple within a row or within a column (and defined above the main diagonal of $\mathbf{P}_\rho$) is to be counted in the total measure of merit. The two options we concentrate on are as follows:

(1) $f(z, y) = \text{sign}(z - y) = +1$ if $z > y$; 0 if $z = y$; and -1 if $z < y$; thus, the (raw) number of satisfactions minus the number of dissatisfactions of the gradient conditions *within rows* above the main diagonal of $\mathbf{P}_\rho$ would be given by the first term in (4.1),

$$\sum_{i<k<j} f(p_{\rho(i)\rho(k)}, p_{\rho(i)\rho(j)}), \qquad (4.2)$$

and the (raw) number of satisfactions minus dissatisfactions of the gradient conditions *within columns* above the main diagonal of $\mathbf{P}_\rho$ would be given by the second term in (4.1),

$$\sum_{i<k<j} f(p_{\rho(k)\rho(j)}, p_{\rho(i)\rho(j)}). \qquad (4.3)$$

To refer to an application of this simple counting mechanism, the phrase *unweighted gradient measure* will be adopted.

(2) $f(z, y) = |z - y|\text{sign}(z - y)$; here, and in contrast to (1), $\text{sign}(z - y)$ is also weighted by the absolute difference between z and y, to generate a *weighted gradient measure* within rows or within columns. Thus, the weighted number of satisfactions minus the number of dissatisfactions of the gradient conditions *within rows* above the main diagonal of $\mathbf{P}_\rho$ would be given by the first term in (4.1) (labeled (4.2) above), and the weighted number of satisfactions minus the number of dissatisfactions of the gradient conditions *within columns* above the main diagonal of $\mathbf{P}_\rho$ would be given by the second term in (4.1) (labeled (4.3) above).

To carry out the GDPP based on the measure in (4.1), an explicit form must be given for the incremental contribution, $M(A_{k-1}, A_k)$, to the total merit measure of patterning generated by placing the single integer in $A_k - A_{k-1}$ at the kth order position. We observe first that the index in (4.1) for the given matrix **P** in its original (identity) order can be rewritten as

$$\sum_{k=1}^{n} I_{row}(k) + \sum_{k=1}^{n} I_{col}(k),$$

where

$$I_{row}(k) \equiv \sum_{i=1}^{k-1} \sum_{j=k+1}^{n} f(p_{ik}, p_{ij})$$

and

$$I_{col}(k) \equiv \sum_{i=1}^{k-1} \sum_{j=k+1}^{n} f(p_{kj}, p_{ij}).$$

Neither $I_{row}(k)$ nor $I_{col}(k)$ depends on the ordering of the objects placed either before or after the index k. Thus, generalizing such a decomposition for any ordering $\rho(\cdot)$ of the rows and columns of **P**, the merit increment for placing an integer, say, k' ($\equiv \rho(k)$) (i.e., $\{k'\} = A_k - A_{k-1}$) at the kth order position can be defined as

$$\sum_{k=1}^{n} I_{row}(\rho(k)) + \sum_{k=1}^{n} I_{col}(\rho(k)),$$

where

$$I_{row}(\rho(k)) = \sum_{i' \in A_{k-1}} \sum_{j' \in S - A_k} f(p_{i'k'}, p_{i'j'}),$$

$$I_{col}(\rho(k)) = \sum_{i' \in A_{k-1}} \sum_{j' \in S - A_k} f(p_{k'j'}, p_{i'j'}),$$

and $A_{k-1} = \{\rho(1), \ldots, \rho(k-1)\}$, $S - A_k = \{\rho(k+1), \ldots, \rho(n)\}$. Thus, letting $\mathcal{F}(A_1) = 0$ for all $A_1 \in \Omega_1$ and using one of the two possible specifications for $f(\cdot, \cdot)$ suggested above (i.e., $f(z,y) = \text{sign}(z-y)$ or $f(z,y) = |z-y|\text{sign}(z-y)$), the recursion in (2.3) can be carried out to identify an optimal row/column reordering of the given proximity matrix **P** to maximize either the unweighted or the weighted gradient measure over all row/column reorderings of **P**.

In using (4.1), *both* the row and column gradient measures in (4.2) and (4.3), respectively, are considered as a sum in optimizing a total merit measure of patterning, where each is defined above the main diagonal of the symmetric reordered matrix $\mathbf{P}_\rho$. It might be noted that either (4.2) or (4.3) could be used by itself and a best reordering of **P** could be constructed that would maximize one or the other through the separate consideration of $\sum_{k=1}^{n} I_{row}(k)$ or $\sum_{k=1}^{n} I_{col}(k)$.[30] Because **P** is symmetric, an optimal row/column reordering based solely on the row gradient measure will also identify an optimal

4.1. OPTIMAL SEQUENCING OF A SINGLE OBJECT SET

row/column reordering based only on the column gradient measure. Specifically, a complete reversal of the row/column reordering that is optimal for the row gradient measure will be an optimal row/column reordering for the column gradient measure (and conversely).[31]

Several numerical applications are given at the end of this section to illustrate the optimal sequencing of a single object set based on optimizing the merit measure in (4.1), as well as just one of its constituent terms in (4.2) or (4.3). For these illustrations, in addition to giving the optimal merit values achieved, convenient descriptive indices are provided for how well the gradient conditions are satisfied. These descriptive ratios are defined by the optimal index values divided by the sum of the contributions for the nonviolations and violations, where the denominators can be interpreted as the maximum the index could be for the given proximities and supposing the gradient conditions were perfectly satisfied. To be explicit, the merit measure in (4.1), for any permutation $\rho(\cdot)$, can be written as the difference between two nonnegative terms, with the first corresponding to the nonviolations of the gradient conditions and the second to the violations. Thus, when the *sum* of these two nonnegative terms is used to divide the difference, the ratio obtained (which is bounded by $+1$ and -1) reflects the (possibly weighted) ratio of the observed index to the maximum that is possible if the gradient conditions were satisfied perfectly. Formally, the descriptive ratio could be given by

$$\frac{\sum_{i<k<j} f(p_{\rho(i)\rho(k)}, p_{\rho(i)\rho(j)}) + \sum_{i<k<j} f(p_{\rho(k)\rho(j)}, p_{\rho(i)\rho(j)})}{\sum_{i<k<j} |f(p_{\rho(i)\rho(k)}, p_{\rho(i)\rho(j)})| + \sum_{i<k<j} |f(p_{\rho(k)\rho(j)}, p_{\rho(i)\rho(j)})|}. \quad (4.4)$$

Analogously, when only the gradient conditions, say, within rows are considered by the use of (4.2), a descriptive ratio could be given by

$$\frac{\sum_{i<k<j} f(p_{\rho(i)\rho(k)}, p_{\rho(i)\rho(j)})}{\sum_{i<k<j} |f(p_{\rho(i)\rho(k)}, p_{\rho(i)\rho(j)})|}. \quad (4.5)$$

As noted, in the numerical applications at the end of this current section, but also in some later generalizations as well, descriptive ratios for an optimal solution will be routinely reported in addition to giving the optimal values achieved for their numerators.

Measures of matrix patterning based on coordinate representations. There are several measures of matrix patterning that can be derived indirectly from the auxiliary problem of attempting to fit a given proximity matrix $\mathbf{P}$ by some type of unidimensional scaling representation. Because detailed discussions of this task are available in the literature (e.g., see Defays (1978); Hubert and Arabie (1986); Groenen (1993); and Hubert, Arabie, and Meulman (1997a)), we merely report the necessary results here and give the derived measure of matrix pattern (used below in the numerical illustrations). Explicitly, suppose we wish to find a set of n ordered coordinate values, $x_1 \leq \cdots \leq x_n$ (such that $\sum_k x_k = 0$), and a permutation $\rho(\cdot)$ to minimize the least-squares criterion

$$\sum_{i<j} (p_{\rho(i)\rho(j)} - |x_j - x_i|)^2.$$

After some algebraic reduction, this latter least-squares criterion can be rewritten as

$$\sum_{i<j} p_{ij}^2 + n \sum_k \left[x_k - \left(\frac{1}{n}\right) G(\rho(k)) \right]^2 - \left(\frac{1}{n}\right) \sum_k [G(\rho(k))]^2,$$

where

$$G(\rho(k)) = \sum_{i=1}^{k-1} p_{\rho(k)\rho(i)} - \sum_{i=k+1}^{n} p_{\rho(k)\rho(i)}.$$

If the measure

$$\sum_{k=1}^{n} [G(\rho(k))]^2 \qquad (4.6)$$

is maximized over all row/column reorderings of $\mathbf{P}$ and the optimal permutation is denoted by $\rho^*(\cdot)$, then $G(\rho^*(1)) \leq \cdots \leq G(\rho^*(n))$, and the optimal coordinates can be retrieved as $x_k = (1/n)G(\rho^*(k))$ for $1 \leq k \leq n$. The minimum value for the least-squares criterion is

$$\sum_{i<j} p_{ij}^2 - \left(\frac{1}{n}\right) \sum_k [G(\rho^*(k))]^2.$$

To execute the GDPP recursion using (4.6), the merit increment for placing the integer, say, k' ($\equiv \rho(k)$) (i.e., $\{k'\} = A_k - A_{k-1}$), in the kth order position can be written as $[G(\rho(k))]^2$, where

$$G(\rho(k)) = \sum_{i' \in A_{k-1}} p_{k'i'} - \sum_{j' \in S-A_k} p_{k'j'},$$

with $A_{k-1} = \{\rho(1),\ldots,\rho(k-1)\}$, $S - A_k = \{\rho(k+1),\ldots,\rho(n)\}$, and $\mathcal{F}(A_1)$ for $A_1 = \{k'\} \in \Omega_1$ defined by

$$\left(\sum_{j' \in S-\{k'\}} p_{k'j'} \right)^2.$$

The recursion in (2.3) can then be carried out to identify an optimal row/column reordering $\rho^*(\cdot)$ of $\mathbf{P}$ based on the measure in (4.6); also, as noted above, the optimal coordinates x_k ($1 \leq k \leq n$) are $(1/n)G(\rho^*(k))$, where the integer $\rho^*(k)$ is placed at the kth order position in an optimal row/column reordering.

To give a more intuitive sense of the measure being optimized in (4.6), we first note that for an optimal row/column reordering $\rho^*(\cdot)$, the variance of the coordinate estimates is maximized (using the fact that the sum of the optimal coordinates, $\sum_k (1/n)G(\rho^*(k))$, is zero). Thus, the pattern of entries sought in an optimally reordered matrix is one in which, within rows, the sum of entries from the left to the main diagonal, versus the sum of entries to the right away from the main diagonal, are maximally separated. The difference between these

4.1. OPTIMAL SEQUENCING OF A SINGLE OBJECT SET

two sums is $G(\rho(k))$, which when squared and aggregated over all rows is the measure in (4.6) that is maximized to identify the optimal permutation $\rho^*(\cdot)$.

As a slight variation on coordinate representation, suppose an equally spaced representation for **P** is to be identified by minimizing the least-squares loss function (over all permutations, $\rho(\cdot)$)

$$\sum_{i<j}(p_{\rho(i)\rho(j)} - \alpha \mid j - i \mid)^2,$$

where α is a multiplicative parameter to be estimated. The measure in (4.6) would be replaced by

$$\sum_k kG(\rho(k)).$$

Thus, the merit increment for placing an integer k' ($\equiv \rho(k)$) (i.e., $\{k'\} = A_k - A_{k-1}$) in the kth order position is $kG(\rho(k))$, and $\mathcal{F}(A_1)$ for $A_1 = \{k'\} \in \Omega_1$ defined by

$$-\sum_{j' \in S - \{k'\}} p_{k'j'}.$$

If we wish, the least-squares estimate for the multiplicative constant α can be obtained as

$$\hat{\alpha} = \sum_{i<j} p_{\rho^*(i)\rho^*(j)} \mid j - i \mid /[n^2(n+1)(n-1)/12],$$

where $\rho^*(\cdot)$ is an optimal row/column reordering of **P**.

Numerical illustrations. To give several examples of sequencing an object set along a continuum based on a symmetric proximity matrix and the various measures of matrix pattern just described, we first derive two separate symmetric proximity matrices from the data given in Table 1.2 on the rated seriousness of thirteen offenses. In particular, for both the 'before' and 'after' proximity matrices, the entry defined for each pair of offenses is the absolute value of the difference in the proportions of rating one offense more serious than the other. (For example, because the proportion judging a bootlegger more serious than a bankrobber is .29 before the movie was shown, a symmetric proximity of $.42 = \mid .29 - .71 \mid$ is given for the pair (bootlegger, bankrobber) in the corresponding proximity matrix.) The two proximity matrices so constructed are given in the upper- and lower-triangular portions of Table 4.1 (where, for graphical convenience, the rows and columns have been already reordered according to the optimal sequencings to be described immediately below).

We first give the optimal orderings associated with an explicit coordinate representation and the measure of matrix pattern given in (4.6).

Optimal orderings based on coordinate estimation, obtained using the program DPSE1U:

Before viewing the movie:

order: 9   7   10   4   2   11   3   5   13   6   12   8   1
coordinates: −.82 −.78 −.33 −.26 −.23 −.17 −.02 .27 .29 .32 .50 .59 .64

Table 4.1: *Symmetric proximity matrices constructed for thirteen offenses using the absolute values of the skew-symmetric proximities generated from the entries in Table 1.2. The above-diagonal entries are before showing the film* Street of Chance; *those below the diagonal are after viewing the motion picture. The entities followed by an asterisk were negative before taking the absolute value.*

offense	9	7	10	4	2	11	3	5	13	6	12	8	1
9:tramp	x	.16	.82	.94	.90	.94	.96	.96	1.0	.98	.98	.97	1.0
7:beggar	.28	x	.72	.98	.86	.92	.96	.96	.98	1.0	1.0	.92	.98
10:speeder	.74	.58	x	.26	.18	.16	.22	.76	.84	.82	.80	.84	.88
4:drunkard	.92	.88	.34	x	.04*	.24	.50	.62	.74	.90	.82	.84	.90
11:petty thief	.90	.88	.28	.06	x	.02*	.42	.52	.62	.84	.80	.84	.86
3:pickpocket	.96	.94	.36	.40	.24	x	.16	.56	.56	.48	.96	.84	.96
2:gambler	.92	.90	.46	.40	.28	.02	x	.34	.36	.50	.74	.72	.84
6:bootlegger	.96	.96	.80	.74	.52	.40	.38	x	.10	.02*	.28	.40	.46
13:smuggler	.96	.96	.78	.68	.52	.40	.38	.02	x	.00	.46	.38	.58
5:quack doctor	.98	.96	.78	.68	.62	.44	.28	.00	.08*	x	.36	.58	.42
12:kidnapper	.98	.96	.84	.84	.94	.68	.46	.40	.28	.30	x	.28	.46*
8:gangster	.98	.98	.88	.78	.78	.74	.64	.46	.32	.36	.24	x	.00
1:bankrobber	1.0	.96	.88	.90	.94	.86	.58	.40	.46	.34	.24*	.00	x

The optimal value for the index in (4.6) is 471.16, with a residual sum-of-squares for the original least-squares task of 3.307 (we might also note that the correlation between the original proximities and the reconstructed absolute values of the coordinate differences is .864. However, this correlation must be considered a conservative association measure, because the optimization process itself did not include the estimation of an additive constant for the proximities, and thus, the correlation was not explicitly optimized).

After viewing the movie:

order: 9   7   10   4   11   3   2   6   13   5   12   8   1
coordinates: −.81 −.75 −.39 −.26 −.21 −.05 .02 .27 .27 .29 .48 .55 .58

Here, the optimal index in (4.6) is 436.71 with a residual sum-of-squares of 2.302 for the original least-squares task (the conservative association measure of .903 was observed for the correlation between the original proximities and the reconstructed absolute values of the coordinate differences).

Both the before and after orderings of the offenses are obviously arranged from least to worst in seriousness, with a clear difference in the position of 'gambler' (which is #2) from the 5th least serious to the 7th, reflecting a change in subjective assessment after viewing the movie. The order of the offenses

4.1. OPTIMAL SEQUENCING OF A SINGLE OBJECT SET

numbered 5 (quack doctor), 6 (bootlegger), and 13 (smuggler) also vary among themselves, but as is apparent in the estimated coordinates both before and after, these three offenses are very close to one another and any interchange from before to after is most likely attributable to minor fluctuations in the data. The same object orders are also optimal for an equally spaced set of coordinates. For both the before and after proximity matrices, the multiplicative constant α was estimated as .12, with, respectively, residual sums of squares of 4.735 and 3.624 and correlations of .759 and .801 obtained between the absolute values of the equally spaced coordinate differences and the proximities.

We give below a brief summary of the optimal orders achieved for the other measures of matrix pattern, i.e., for the weighted or unweighted gradients within rows (or columns) alone or within both rows and columns. In each case the optimal index value is given and separated into the constituent positive and negative contributions, depending on the nonviolations and violations for the achieved optimal order. Also, as given by the terms in (4.4) and (4.5) and defined by a ratio, a descriptive index is provided in each case for how well the gradient condition is satisfied.

Optimal orderings based on other measures of matrix pattern, obtained using DPSE1U:

Before viewing the movie:

Criterion	Index	Order
within row (or column) unweighted gradient	236 = (256−20) (ratio = .855)	9 7 4 10 2 11 3 13 6 5 12 8 1
within row and column unweighted gradient	431 = (489−58) (ratio = .788)	9 7 4 10 2 11 3 5 13 6 12 8 1
within row (or column) weighted gradient	95.48 = (97.56−2.08) (ratio = .958)	9 7 10 4 2 11 3 13 5 6 12 8 1
within row and column weighted gradient	161.45 = (165.66−4.21) (ratio = .950)	9 7 10 4 2 11 3 5 13 6 12 8 1

After viewing the movie:

Criterion	Index	Order
within row (or column) unweighted gradient	263 = (272−9) (ratio = .936)	9 7 10 4 11 3 2 6 5 13 12 8 1
within row and column unweighted gradient	491 = (515−24) (ratio = .911)	9 7 10 4 11 3 2 13 6 5 12 8 1
within row (or column) weighted gradient	104.44 = (105.00−.56) (ratio = .989)	9 7 10 4 11 3 2 6 13 5 12 8 1
within row and column weighted gradient	165.08 = (166.56−1.48) (ratio = .982)	9 7 10 4 11 3 2 6 13 5 12 8 1

These latter results are very consistent with those generated from the coordinate representations — the offense 'gambler' is 5th least serious in all orders for the before matrix and 7th for the after matrix; there is also some of the same unsystematic interchange among the 'close' offenses 5, 6, and 13, and two instances of an interchange of the offenses 4 (drunkard) and 10 (speeder) when using the unweighted gradient measure for the before matrix. We might also observe that for the optimal indices and their descriptive ratios calculated using (4.4) and (4.5), the after matrix is apparently structured slightly better than the before matrix (this difference was also reflected in the lower residual sums of squares and higher correlations provided in the context of coordinate estimation).

4.1.2 Skew-Symmetric One-Mode Proximity Matrices

As noted earlier, any skew-symmetric matrix $\mathbf{P}^{SS} = \{p_{ij}^{SS}\}$ contains two distinct types of information about the relationship between any pair of objects. First, for each object pair, $O_i, O_j \in S$, $|p_{ij}^{SS}|$ indicates a (symmetric) degree of dissimilarity between O_i and O_j, whereas $\text{sign}(p_{ij}^{SS})$ indicates the directionality of dominance. Thus, if p_{ij}^{SS} is positive, O_i can be interpreted as dominating O_j with the magnitude of p_{ij}^{SS} reflecting the degree of dominance. In the presence

4.1. OPTIMAL SEQUENCING OF A SINGLE OBJECT SET

of a skew-symmetric matrix $\mathbf{P}^{SS}$, a natural extension of the notion of an anti-Robinson form appropriate for a symmetric proximity matrix $\mathbf{P}$ would be to a row/column reordering of $\mathbf{P}^{SS}$, say, by a permutation $\rho(\cdot)$, in which (a) the degree of dominance is perfectly depicted by having an anti-Robinson form in the reordered matrix $\{|\ p^{SS}_{\rho(i)\rho(j)}\ |\}$, and (b) using the *same* row/column reordering, the direction of dominance is perfectly depicted in the reordered matrix $\{\text{sign}(p_{\rho(i)\rho(j)})\}$ in that all above-diagonal entries are nonnegative and all below-diagonal entries are (therefore) nonpositive.

The various classes of measures appropriate for indexing matrix patterning and for obtaining an optimal reordering of $\mathbf{P}^{SS}$ could rely on either $\{|\ p^{SS}_{ij}\ |\}$ or $\{\text{sign}(p^{SS}_{ij})\}$ alone or, instead, attempt to use both types of information jointly by considering the matrix $\mathbf{P}^{SS}$ as is. (Thus, there is the possibility of assessing the extent to which the same [or similar] optimal reorderings arise from the use of the separate information sources.) If $\{|\ p^{SS}_{ij}\ |\}$ were considered by itself, the measures described in the previous section for symmetric proximities would obviously be appropriate; in fact, the numerical examples given in the last section for sequencing a symmetric one-mode proximity matrix relied on dissimilarity matrices of this latter form constructed from the nonsymmetric data of Table 1.2. The more distinctive class of alternatives to be considered below is best viewed as emphasizing $\{\text{sign}(p^{SS}_{ij})\}$ and forcing the above-diagonal entries in an optimally reordered matrix to be as consistently positive as possible.

More pointedly, we will *not* discuss at this juncture the optimization through row/column reorderings of weighted or unweighted gradient measures for the matrix $\mathbf{P}^{SS}$ that would attempt to force an anti-Robinson form for the above-diagonal entries in the reordered matrix and which would parallel those developed for symmetric matrices. The authors' experience has found that these *may* work well, but anomalies can arise frequently. For example, a row/column reordering of $\mathbf{P}^{SS}$ may exist in which the reordered matrix $\{|\ p^{SS}_{\rho(i)\rho(j)}\ |\}$ is very close to being anti-Robinson in form, and for this same reordering the above-diagonal entries in $\{p^{SS}_{\rho(i)\rho(j)}\}$ tend to be positive, but an optimal reordering of the matrix $\mathbf{P}^{SS}$ based on the above-diagonal gradient conditions analogous to those used for symmetric matrices finds an even better reordering, where the above-diagonal entries may no longer be predominantly positive. Although formally better according to the particular gradient index chosen, the latter reordering is also hard to explain in any clear substantive fashion. Therefore, the use of gradient conditions for the matrix $\mathbf{P}^{SS}$, as is, may be best avoided when we might want to identify an anti-Robinson form for the mostly positive above-diagonal entries in a reordered matrix. We note, in a discussion at the end of this section, certain kinds of gradient conditions we might want to identify in a reordered matrix $\{p^{SS}_{\rho(i)\rho(j)}\}$, but in an unfolding context where the original nonsymmetric matrix (before generating the skew-symmetric proximity matrix $\mathbf{P}^{SS}$) is used. Here, both subjects and objects are assumed orderable along the same continuum and subjects generally make their preference judgments as a function of their own (estimated) distances to the objects.

Given $\{p_{ij}^{SS}\}$ (or possibly just $\{\text{sign}(p_{ij}^{SS})\}$), an obvious class of measures of a matrix pattern would be the sum of the above-diagonal entries

$$\sum_{i<j} p_{ij}^{SS} \quad \left(\text{or} \sum_{i<j} \text{sign}\left(p_{ij}^{SS}\right)\right).$$

In fact, because $p_{ij}^{SS} = \text{sign}(p_{ij}^{SS}) \mid p_{ij}^{SS} \mid$, the index $\sum_{i<j} p_{ij}^{SS}$ can be interpreted merely as a weighted version of the one based just on $\text{sign}(p_{ij}^{SS})$. We will assume in our discussion below that $\sum_{i<j} p_{ij}^{SS}$ is being considered, but the obvious replacement of p_{ij}^{SS} by $\text{sign}(p_{ij}^{SS})$ could incorporate the use of $\sum_{i<j} \text{sign}(p_{ij}^{SS})$ directly.

To carry out the GDPP based on $\sum_{i<j} p_{ij}^{SS}$, the incremental contribution to the total merit measure of patterning generated by placing the single integer, say, $k' (\equiv \rho(k))$ in $A_k - A_{k-1}$, at the kth order position can be defined as

$$J(\rho(k)) \equiv \sum_{j' \in A_{k-1}} p_{j'k'}^{SS}. \tag{4.7}$$

Letting $\mathcal{F}(A_1) = 0$ for $A_1 \in \Omega_1$, the recursion in (2.3) can be executed to identify an optimal row/column reordering of $\mathbf{P}^{SS}$ that will maximize the sum of the above-diagonal entries over all row/column reorderings of $\mathbf{P}^{SS}$.

The optimization task of reordering the rows/columns of a matrix to maximize the sum of above-diagonal entries and its solution by the type of DP strategy sketched above was first introduced by Lawler (1964) to identify minimum feedback arc sets in a directed graph. The optimization task itself, however, has several other distinct substantive incarnations, e.g., in maximum likelihood paired comparison ranking (Flueck and Korsh (1974)) or in triangulating an input-output matrix (Korte and Oberhofer (1971)). For an extensive review of the variety of possible applications up to the middle 1970's, the reader is referred to Hubert (1976). For more up-to-date surveys and current work, see Charon, Hudry, and Woirgard (1996), Barthélemy et al. (1995), and Charon et al. (1997).

Measures of matrix pattern for a skew-symmetric matrix $\mathbf{P}^{SS}$ that might be derivable indirectly from an attempt to generate some type of coordinate representation have a rather different status than they had for a symmetric matrix $\mathbf{P}$. In the skew-symmetric framework, closed-form least-squares solutions are possible, thus eliminating the need for any GDPP optimization strategy. For example, suppose we wish to find a set of n coordinate values $x_1, \ldots, x_n$ such that the least-squares criterion,

$$\sum_{i<j}(p_{ij}^{SS} - (x_j - x_i))^2,$$

is minimized. An optimal set of coordinates can be obtained analytically (e.g., see Hubert and Arabie (1986)) by letting x_j be the average proximity within

4.1. OPTIMAL SEQUENCING OF A SINGLE OBJECT SET

column j, i.e., $x_j = (1/n) \sum_i p_{ij}^{SS}$, with a minimum least-squares loss value of $\sum_{i<j} (p_{ij}^{SS})^2 - n \sum_j (\sum_i p_{ij}^{SS}/n)^2$. Thus, an optimal row/column reordering of $\mathbf{P}^{SS}$ can be obtained merely by using the order of the optimal coordinates from smallest (most negative) to largest (most positive). Similarly, if we consider an equally spaced coordinate representation obtained by minimizing the least-squares loss function

$$\sum_{i<j} (p_{ij}^{SS} - \alpha(x_j - x_i))^2,$$

where $x_1, \ldots, x_n$ are the integers $1, \ldots, n$ in some order and α is some multiplicative constant to be estimated, the optimal row/column reordering of $\mathbf{P}^{SS}$ induced by the integer coordinates would again be generated by the ordering of $(1/n) \sum_i p_{ij}^{SS}$ for $1 \leq j \leq n$. In the numerical examples given below, these latter analytic solutions are illustrated in addition to those measures described earlier defined by an above-diagonal sum. Because no analytic solution is possible when using these latter measures, a GDPP application is required for their optimization.

Numerical illustration. To provide a few examples of sequencing an object set along a continuum based on a skew-symmetric proximity matrix, we reconsider the data of Table 1.2 and first form two skew-symmetric matrices (one associated with before showing the motion picture and one after) based on the signed differences between the proportions of rating one offense as more serious than the other. The absolute values of these differences were given in Table 4.1 and used to illustrate the sequencing of object sets based on symmetric proximity information. We merely indicate by asterisks beside the relevant entries in Table 4.1 which differences were negative before the absolute values were taken; thus, with this annotation, both the before and after skew-symmetric matrices can be considered displayed in Table 4.1 as well (although using the object reorderings obtained from coordinate representations derived from their absolute values).

The optimal orderings based on maximizing the sum of above-diagonal entries for the two skew-symmetric proximity matrices are as follows (where the offense of being a gambler, #2, is bracketed for emphasis):

before: 9 7 10 [2] 4 11 3 6 5 13 1 12 8;

after: 9 7 10 4 11 3 [2] 6 5 13 1 12 8.

The before order given above produces a reordered skew-symmetric matrix in which all the entries above the main diagonal are nonnegative except the (2,11) pair, which has a value of $-.02$; the after order produces a reordered skew-symmetric matrix where all entries above the main diagonal are nonnegative without exception. In comparison with the orderings based on the symmetric proximities of Table 4.1, there are a few minor local interchanges among (2,4), (5,6,13), and (1,8,12) that force a little more nonnegativity above the main diagonals (in comparison with the orderings used in presenting Table 4.1), but the results are otherwise very consistent. Based on the optimal orderings for the skew-symmetric information, there is an obvious change in the position of 'gambler' (i.e., offense #2) from the 4th least serious before the showing of the movie to the 7th thereafter. We also note that these same two before and after optimal reorderings (in the present case) are also optimal if only the sign information from the original skew-symmetric proximity matrices is considered.

For completeness we give below the object sequences obtained from closed-form coordinate representations based on the before and after skew-symmetric proximity matrices along with some associated descriptive information. Again, there is an expected consistency with respect to the orderings obtained by maximizing the sum of above-diagonal entries with some minor local interchange differences among (2,4), (5,6,13), and (1,8,12).

Optimal orderings based on coordinate estimation for the skew-symmetric proximity matrices, obtained using DPSE1U; the residual sum-of-squares and the correlations between the skew-symmetric proximities and the signed coordinate differences are also provided:

before: 9 7 10 4 2 11 3 5 13 6 12 1 8;
coordinates: -.82 -.78 -.33 -.26 -.23 -.18 -.02 .27 .29 .32 .57 .57 .59;

before: equally spaced coordinates ($\hat{\alpha} = .12$): 5.182; correlation = .739;
   estimated coordinates: 3.457; correlation = .852;

after: 9 7 10 4 11 3 2 6 5 13 12 1 8;
coordinates: -.81 -.75 -.39 -.26 -.21 -.05 .02 .27 .28 .29 .51 .55 .55.

after: equally spaced coordinates ($\hat{\alpha} = .12$): 3.853; correlation = .788;
   estimated coordinates: 2.369; correlation = .896.

The Greenberg Form:

As noted earlier in this section, there is at least one specific context in which it is of interest to identify whether the matrix $\mathbf{P}^{SS}$, as is, may be row/column reorderable to satisfy a particular set of gradient conditions within its rows and columns. Explicitly, suppose the set S of n objects is assumed orderable along a continuum and, in addition, a group of judges can be placed along the same continuum so that when asked which of two presented objects they would prefer, the response is made according to the closer distance between the judge's location and the location of the two objects presented. If the skew-symmetric matrix $\mathbf{P}^{SS} = \{p_{ij}^{SS}\}$ is constructed by letting p_{ij}^{SS} = the proportion of judges preferring object O_i to O_j − the proportion of judges preferring object O_j to

4.1. OPTIMAL SEQUENCING OF A SINGLE OBJECT SET

O_i, then when $\mathbf{P}^{SS}$ is row/column reordered according to the placement of the objects along the continuum, say, as $\{p^{SS}_{\rho(i)\rho(j)}\}$, the specific set of gradient conditions we give below will be satisfied (these were first observed and discussed by Greenberg (1965) and will be referred to as defining a Greenberg form for the matrix $\{p^{SS}_{\rho(i)\rho(j)}\}$):

within rows: $p^{SS}_{\rho(i)\rho(k)} \leq p^{SS}_{\rho(i)\rho(j)}$ for $1 \leq i < k < j \leq n$;
within columns: $p^{SS}_{\rho(k)\rho(j)} \geq p^{SS}_{\rho(i)\rho(j)}$ for $1 \leq i < k < j \leq n$.

In words, the reordered matrix $\mathbf{P}^{SS}_\rho = \{p^{SS}_{\rho(i)\rho(j)}\}$ has a Greenberg form if the entries within a row moving to the right from the main diagonal never decrease, and if the entries within a column moving up from the main diagonal never increase. (In comparison to the gradient conditions for an anti-Robinson form for a symmetric matrix $\mathbf{P}$, the obvious difference here is in the reversal of the within-column inequality when identifying a Greenberg form for a skew-symmetric matrix $\mathbf{P}^{SS}$.)

The natural merit measures of how well a particular reordered skew-symmetric proximity matrix satisfies the Greenberg gradient conditions, which could be used in the GDPP to order the objects along the continuum, would be the analogue of (4.1),

$$\sum_{i<k<j} f(p^{SS}_{\rho(i)\rho(k)}, p^{SS}_{\rho(i)\rho(j)}) + \sum_{i<k<j} f(p^{SS}_{\rho(i)\rho(j)}, p^{SS}_{\rho(k)\rho(j)}),$$

where, in contrast to (4.1), the second within-column comparison term, $\sum_{i<k<j} f(p^{SS}_{\rho(i)\rho(j)}, p^{SS}_{\rho(k)\rho(j)})$, has an interchange of its two components.[32]

Numerical illustrations. To give an example of the use of the Greenberg gradient conditions in optimally ordering a set of objects along a continuum, we use a data set reported in Orth (1989) on the preferences for five German political parties collected in 1980. The five parties are ordered below according to their position on a political left-right dimension:

DKP: Deutsche Kommunistische Partei (German Communist Party),
SPD: Sozialdemokratische Partei Deutschlands (Social Democratic Party),
FDP: Freie Demokratische Partei (Free Democratic Party),
CDU/CSU: Christliche Demokratische Union/Christliche Soziale Union
  (Christian Democratic Union/Christian Social Union),
NPD: Nationaldemokratische Partei Deutschlands
  (National Democratic Party).

Based on the complete preference rankings for 1316 German voters, Table 4.2 provides a 5×5 skew-symmetric matrix for the five political parties derived from the choice proportions.

The optimal orderings for the matrix of Table 4.2 for both the weighted and unweighted gradient measures are given below, along with their indices of merit

Table 4.2: *A skew-symmetric proximity matrix among five German political parties, obtained from the complete preference rankings for 1316 German voters.*

party	DKP	SPD	FDP	CDU/CSU	NPD
DKP	x	-.982	-.968	-.862	-.015
SPD	.982	x	.181	.131	.933
FDP	.968	-.181	x	.114	.964
CDU/CSU	.862	-.131	-.114	x	.945
NPD	.015	-.933	-.964	-.945	x

and descriptive ratios defined by the direct extension of equation (4.4) to the Greenberg gradient conditions. In addition, the indices of merit are given as if the parties were merely ordered, as discussed by Orth (1989), according to their position on the political left-right dimension.

unweighted gradient:

    optimal order: DKP → SPD → CDU/CSU → FDP → NPD
    index = 16 = (18−2)
    ratio = .800 = (18−2)/(18+2)

    left-right order: DKP → SPD → FDP → CDU/CSU → NPD
    index = 14 = (17−3)
    ratio = .700 = (17−3)/(17+3)

weighted gradient:

    optimal order: DKP → FDP → SPD → CDU/CSU → NPD
    index = 11.707 = (11.771−.064)
    ratio = .989 = (11.771−.064)/(11.771+.064)

    left-right order: DKP → SPD → FDP → CDU/CSU → NPD
    index = 11.683 = (11.769−.086)
    ratio = .987 = (11.769−.086)/(11.769+.086)

Obviously, for either the weighted or unweighted gradient measures, the left-right order is *not* optimal and better orderings can be identified. This observation casts doubt on the reasonableness of the initial assumption that voters and political parties are jointly orderable along a common left-right continuum and that voters would prefer those parties that are closer to their own locations. The two extreme parties, DKP and NPD, are placed appropriately at the ends of the continuum in the optimal orderings given above, but there is difficulty with the three intermediates, SPD, FDP, and CDU/CSU. Orth (1989) addresses this point explicitly and notes that because coalition governments involving SPD, FDP, and CDU/CSU have been the norm for the last several decades, consistent preferences based on a clear-cut left-right continuum are probably inherently difficult to observe for these intermediate parties.

4.1. OPTIMAL SEQUENCING OF A SINGLE OBJECT SET

Besides attempting to obtain an ordering of the parties along a continuum that would induce an approximate Greenberg form within the rows and columns of the reordered matrix, it is also possible to reorder the matrix of Table 4.2 using the earlier criterion of maximizing the sum of above-diagonal entries. Doing so is an attempt to define a dominance ordering among the political parties. An optimal reordering of Table 4.2 maximizing the above-diagonal sum is SPD → FDP → CDU/CSU → NPD → DKP (the maximum sum is .610, and all above-diagonal entries are nonnegative).[33] This specific ordering can be interpreted as a consensus (or societal) ordering for the political parties over the set of judges.

We might note in closing this illustration that there is an extensive literature on the use of binary choice proportions in obtaining a transitive societal ordering for a set of objects (that form the items of choice) by maximizing the above-diagonal sum in a reordered proximity matrix. The reader is referred to Bowman and Colantoni (1973; 1974) or the review by Hubert (1976) for many more details.

4.1.3 Two-Mode Proximity Matrices

As noted in the introduction to Chapter 4, if the available proximity data are between two distinct sets, S_A and S_B, and in the form of an $n_A \times n_B$ nonnegative dissimilarity matrix $\mathbf{Q}$, a symmetric proximity matrix $\mathbf{P}^{AB}$ can be constructed for the single object set $S = S_A \cup S_B$ ($n = n_A + n_B$) that will have missing entries for object pairs within S_A and within S_B. An optimal joint sequencing of $S_A \cup S_B$ along a continuum can then be attempted using some measure of matrix patterning defined for $\mathbf{P}^{AB}$, with the obvious candidates being the same as those considered in Section 4.1.1. Either the weighted or the unweighted gradient measures of (4.1) based on the function $f(\cdot, \cdot)$ extend directly to a use with $\mathbf{P}^{AB}$ merely by defining $f(z, y)$ to be identically 0 if either z or y refers to a pairwise proximity within S_A or within S_B (and thus, either z or y refers to a missing proximity); applications of these measures will be given in the numerical examples. In contrast, however, the obvious analogues for the measures of matrix patterning based on coordinate representation fail to provide a way of constructing merit increments that are independent of the order of the objects previously placed (as was briefly noted in the introduction to this monograph), and thus cannot be directly implemented within a GDPP framework. We will return to this topic at the end of this section and give a more detailed explanation of why this difficulty arises.

Analogous to the case of complete symmetric proximity matrices $\mathbf{P}$, the use of the weighted or unweighted gradient measure to reorder $\mathbf{P}^{AB}$ optimally can be interpreted as an attempt to find a row/column reordering as close as possible to an anti-Robinson form in its nonmissing entries. In turn, any reordering of $\mathbf{P}^{AB}$ must also induce separate row and column reorderings for the original $n_A \times n_B$ proximity matrix $\mathbf{Q}$. If $\mathbf{P}^{AB}$ can be placed in a perfect anti-Robinson form (as, for example, when a one-dimensional Euclidean representation exists for the objects in $S = S_A \cup S_B$, or a perfect ultrametric defines the nonmissing entries in $\mathbf{P}^{AB}$), the reordered matrix $\mathbf{Q}$ would also display a perfect order pattern for its entries within each row and within each column. Explicitly, the entries within each row (or column) would be nonincreasing to a minimum and nondecreasing thereafter. Such a pattern can be called a Q-form (within rows and within columns) in the reordered matrix $\mathbf{Q}$ (as named by Kendall (1971a; 1971b)), and has a very long history in the literature of unidimensional unfolding (e.g., see Coombs (1964), Chapter 4; Hubert and Arabie (1995a)).

In the application of the weighted or unweighted gradient conditions for reordering a symmetric matrix $\mathbf{P}$, one possible variation discussed was to allow the gradient measures (defined above the main diagonal) to be optimized only within rows (or equivalently, only within columns). For two-mode data and the derived matrix $\mathbf{P}^{AB}$, a different restriction on gradient comparisons may also be useful. Specifically, suppose that the gradient conditions are used to define an optimal reordering of $\mathbf{P}^{AB}$ within both the rows and columns of $\mathbf{P}^{AB}$ but only for those in which the common object on which a gradient comparison is made is a member of S_B (containing those objects forming the columns of $\mathbf{Q}$). This approach can be carried out directly in the use of the gradient measures in (4.1) merely by defining $f(z, y)$ to be identically zero whenever the common object associated with the two proximities z and y is not a member of S_B. If this restricted measure is adopted as the means to reorder $\mathbf{P}^{AB}$ optimally, and if there are no violations of these gradient conditions in the optimally reordered matrix $\mathbf{P}^{AB}$, then the induced reordering of the $n_A \times n_B$ matrix $\mathbf{Q}$ would have a perfect Q-form within each column (but not necessarily within each row). More generally, the attempt to reorder $\mathbf{P}^{AB}$ optimally based on this restricted measure can be interpreted as a mechanism for reordering the rows of $\mathbf{Q}$ to approximate a Q-form within columns (alone). Analogously, if the function $f(z, y)$ is defined to be identically zero when the common object associated with the proximities z and y is not a member of S_A, then an optimal reordering of $\mathbf{P}^{AB}$ can be interpreted as a mechanism for reordering the columns of $\mathbf{Q}$ to approximate a Q-form within rows (alone).

In addition to these possible variations on how the weighted or unweighted gradient conditions can be evaluated in a reordered matrix, there is one additional alternative for the choice of an index for matrix patterning (although for brevity, we will not explicitly illustrate it in our numerical examples below). Instead of maximizing the weighted or unweighted gradient conditions as they have been defined, we can also allow the minimization of just the violations (either weighted or unweighted). For some data sets, slightly different optimal joint orderings of S_A and S_B might arise when emphasizing the latter measure because it relies only on the reduction of gradient violations and not on the maximization of a difference between the gradient nonviolations and violations.[34]

Numerical illustrations. As a numerical example of optimally reordering a two-mode proximity matrix $\mathbf{Q}$, we will use the data of Table 1.3 on the dissimilarities between the goldfish retinal receptors (the eleven rows of $\mathbf{Q}$) and

4.1. OPTIMAL SEQUENCING OF A SINGLE OBJECT SET

the specific wavelengths of light (the nine columns of $\mathbf{Q}$). The joint optimal sequencings of the rows/columns of the derived matrix $\mathbf{P}^{AB}$ are given below for the weighted and unweighted gradient measures, along with the relevant descriptive information (the column objects are shown by an underline in each joint ordering).

Optimal orderings for the combined row/column object set based on the gradient measures, obtained using DPSE2U:

within row and column unweighted gradient:
joint order: $\underline{3}$ 10 11 5 7 $\underline{8}$ 2 1 4 6 2 $\underline{7}$ 1 $\underline{5}$ $\underline{6}$ 8 9 3 $\underline{4}$ $\underline{9}$
row order: 10 11 5 7 1 4 6 2 8 9 3
column order: $\underline{3}$ $\underline{8}$ $\underline{2}$ $\underline{7}$ $\underline{1}$ $\underline{5}$ $\underline{6}$ $\underline{4}$ $\underline{9}$
index = 464 = (517−53); ratio = .814 = (517−53)/(517+53)

within row and column weighted gradient:
joint order: $\underline{3}$ $\underline{8}$ 10 11 $\underline{2}$ 5 7 1 $\underline{7}$ 1 4 6 2 $\underline{5}$ $\underline{6}$ 9 9 $\underline{4}$ 8 3
row order: 10 11 5 7 1 4 6 2 9 8 3
column order: $\underline{3}$ $\underline{8}$ $\underline{2}$ $\underline{7}$ $\underline{1}$ $\underline{5}$ $\underline{6}$ $\underline{9}$ $\underline{4}$
index = 31025 = (32019−994); ratio = .940 = (32019−994)/(32019+994)

Although there is some variation in how the row and column objects are combined for the weighted and unweighted gradient measures in the joint orders given above, the separate row and column reorderings are very consistent, with one adjacent receptor interchange for 8 and 9 and one adjacent wavelength interchange for $\underline{4}$ and $\underline{9}$. The column ordering of the stimuli for the unweighted gradient measure (i.e., $\underline{3}$ $\underline{8}$ $\underline{2}$ $\underline{7}$ $\underline{1}$ $\underline{5}$ $\underline{6}$ $\underline{4}$ $\underline{9}$) is exactly consistent with decreasing wavelength (for the weighted gradient measure, the orders of the two lowest wavelength stimuli of $\underline{4}$ and $\underline{9}$ are reversed but otherwise the order is the same as for the unweighted gradient measure).

As noted in our earlier discussion, we can allow some variation in how the gradient measures are obtained, particularly when allowing the comparisons to be defined only within the row or only within the column objects of the original two-mode matrix $\mathbf{Q}$. For example, if gradient comparisons are limited to within the rows (the receptors) of $\mathbf{Q}$ to emphasize an approximate Q-form over columns (the light stimuli), the optimal orderings of the eight wavelength stimuli for both the unweighted and weighted gradient measures turn out to be identical to those obtained for the unweighted gradient measure within both rows and columns (i.e., $\underline{3}$ $\underline{8}$ $\underline{2}$ $\underline{7}$ $\underline{1}$ $\underline{5}$ $\underline{6}$ $\underline{4}$ $\underline{9}$) but for obviously different optimal index values (i.e., within-row unweighted gradient: index = 264 = (289−25); within-row weighted gradient: index = 19293 = (19926−633)).

In closing this section, we will illustrate the difficulties encountered in a two-mode context when we attempt to derive a measure of matrix pattern based on a unidimensional coordinate representation that could then be applied in the GDPP framework (which would be intended to parallel the measure in (4.6) for a one-mode symmetric proximity matrix $\mathbf{P}$). The two-mode coordinate representation task can be phrased as follows: given the $n \times n$ matrix $\mathbf{P}^{AB} =$

$\{p_{ij}^{AB}\}$ (where for computational convenience the asterisk entries denoting the missing values can be considered replaced by zeros) and an $n \times n$ indicator matrix $\mathbf{W} = \{w_{ij}\}$, where $w_{ij} = 0$ for $1 \leq i, j \leq n_A$; $n_A + 1 \leq i, j \leq n$, and $= 1$ otherwise, find a set of n ordered coordinate values, $x_1 \leq \cdots \leq x_n$ (such that $\sum_k x_k = 0$), minimizing the least-squares criterion

$$\sum_{i<j} w_{\rho(i)\rho(j)} \left(p_{\rho(i)\rho(j)}^{AB} - |x_j - x_i| \right)^2.$$

(Obviously, the purpose of the indicator value $w_{\rho(i)\rho(j)}$ is to choose only squared discrepancies that involve the nonmissing proximities in $\mathbf{P}^{AB}$.) Relying on the extensive work of Heiser (1981), Chapter 6, a solution to the (necessary) stationary equations derived for the least-squares loss function would produce a set of coordinates $x_1 \leq \cdots \leq x_n$ (where $\sum_k x_k = 0$) and an associated permutation $\rho(\cdot)$ of the n objects in S $(= S_A \cup S_B)$, such that if we let

$$K(\rho(k)) = \sum_{i=1}^{k-1} w_{\rho(k)\rho(i)} p_{\rho(k)\rho(i)} - \sum_{i=k+1}^{n} w_{\rho(k)\rho(i)} p_{\rho(k)\rho(i)},$$

and if $\rho(k)$ is a row object in S_A, then

$$x_k = \frac{1}{n_B} \left[K(\rho(k)) - \frac{1}{n_A + n_B} \sum_{\rho(j) \in S_A} K(\rho(k)) \right];$$

also, if $\rho(k)$ is a column object in S_B, then

$$x_k = \frac{1}{n_A} \left[K(\rho(k)) - \frac{1}{n_A + n_B} \sum_{\rho(j) \in S_B} K(\rho(k)) \right].$$

Now, the least-squares loss function for any permutation $\rho(\cdot)$ can be rewritten as

$$\sum_{i<j}(p_{ij}^{AB})^2 + n_B \sum_{\rho(k) \in S_A} \left[x_k - \frac{1}{n_B} K(\rho(k)) + \frac{1}{n_A + n_B} \sum_{\rho(j) \in S_A} K(\rho(j)) \right]^2$$

$$+ n_A \sum_{\rho(k) \in S_B} \left[x_k - \frac{1}{n_A} K(\rho(k)) + \frac{1}{n_A + n_B} \sum_{\rho(j) \in S_B} K(\rho(j)) \right]^2$$

$$-2 \left(\sum_{\rho(k) \in S_A} x_k - \sum_{\rho(k) \in S_A} \left[\frac{1}{n_B} K(\rho(k)) - \frac{1}{n_A + n_B} \sum_{\rho(j) \in S_A} K(\rho(j)) \right] \right)$$

$$\times \left(\sum_{\rho(k) \in S_B} x_k - \sum_{\rho(k) \in S_B} \left[\frac{1}{n_A} K(\rho(k)) - \frac{1}{n_A + n_B} \sum_{\rho(j) \in S_B} K(\rho(j)) \right] \right)$$

4.1. OPTIMAL SEQUENCING OF A SINGLE OBJECT SET

$$-\left[\frac{1}{n_B}\sum_{\rho(k)\in S_A}(K(\rho(k)))^2 + \frac{1}{n_A}\sum_{\rho(k)\in S_B}(K(\rho(k)))^2\right.$$
$$\left.+\frac{1}{n_A n_B}\sum_{\rho(k)\in S_A}K(\rho(k))\sum_{\rho(k)\in S_B}K(\rho(k))\right].$$

Thus, if $\rho(\cdot)$ is derived from a stationary solution, all the nonconstant terms are zero except the last three in the equation above, i.e.,

$$\frac{1}{n_B}\sum_{\rho(k)\in S_A}(K(\rho(k)))^2 + \frac{1}{n_A}\sum_{\rho(k)\in S_B}(K(\rho(k)))^2$$
$$+\frac{1}{n_A n_B}\sum_{\rho(k)\in S_A}K(\rho(k))\sum_{\rho(k)\in S_B}K(\rho(k)). \quad (4.8)$$

Analogous to results from using a symmetric proximity matrix with (4.6), if one could show that maximizing (4.8) leads to a stationary solution and if (4.8) could produce additive merit increments (again, analogous to the use of (4.6)), a GDPP recursive solution could be generated. Unfortunately, it does not appear possible to develop such additive merit increments based on (4.8) because of the presence of the last term

$$\frac{1}{n_A n_B}\sum_{\rho(k)\in S_A}K(\rho(k))\sum_{\rho(k)\in S_B}K(\rho(k)),$$

which requires knowledge of how an entity in A_{k-1} was reached (i.e., a knowledge of the values of $K(\rho(k))$ for those objects $\rho(k)\in A_{k-1}$), and the latter depends on the order of object placement in the construction of A_{k-1}. If only the first two terms in (4.8) were present, a GDPP recursive process could be carried out, but with the inclusion of the third, it appears that an obvious GDPP specialization is not possible.

4.1.4 Object Sequencing for Symmetric One-Mode Proximity Matrices Based on the Construction of Optimal Paths

The type of GDPP recursive process for the optimal sequencing of an object set introduced in Section 4.1 (and implemented in the preceding sections, 4.1.1 through 4.1.3) is based on a definition for the sets $\Omega_k, 1\leq k\leq n$, characterized by all subsets containing k of the subscripts on the objects in S. There is, however, at least one other possibility for how these basic sets might be redefined and how a recursive process might then be carried out, which would allow a different measure of matrix patterning to be used in optimally reordering a symmetric one-mode proximity matrix $\mathbf{P}$. The specific variation discussed here involves the construction of a sequencing of the objects in S by identifying an optimal path between the objects (that includes each object exactly once) based on some function of the proximities between adjacently-placed objects (for example, we might minimize or maximize either the sum of such adjacent proximities or their maximum or minimum). The matrix pattern being assessed in the reordered matrix $\mathbf{P}_\rho$ would involve the magnitudes of those proximities immediately adjacent to the principal diagonal, i.e., $p_{\rho(i)\rho(i+1)}$ for $1 \leq i \leq n-1$. For example, in one manifestation of the various optimization options possible, we may wish to minimize the sum

$$\sum_{i=1}^{n-1} p_{\rho(i)\rho(i+1)}$$

over all possible reorderings of $\mathbf{P}$.

To be explicit in the required tailoring of the GDPP (and for the moment emphasizing the minimization of the sum of adjacent object proximities in constructing a path among the objects in S), a collection of sets $\Omega_1, \ldots, \Omega_n$ is defined (thus again, $K \equiv n$) so that each entity in $\Omega_k, 1 \leq k \leq n$, is now an ordered pair (A_k, j_k). Here, A_k is a k-element subset of the n subscripts on the objects in S, and j_k is one subscript in A_k (to be interpreted as the subscript for the last-placed object in a sequencing of the objects contained within A_k). The function value $\mathcal{F}((A_k, j_k))$ is the optimal contribution to the total measure of matrix patterning for the objects in A_k when they are placed in the first k positions in the (re)ordering, and when the object with subscript j_k occupies the kth. A transformation is possible between $(A_{k-1}, j_{k-1}) \in \Omega_{k-1}$ and $(A_k, j_k) \in \Omega_k$ if $A_{k-1} \subset A_k$ and $A_k - A_{k-1} = \{j_k\}$ (i.e., A_{k-1} and A_k differ by the one integer j_k). The cost increment $C((A_{k-1}, j_{k-1}), (A_k, j_k))$ is simply $p_{(j_{k-1})j_k}$ for the contribution to the total measure of patterning generated by placing the object with the single integer subscript in $A_k - A_{k-1}$ at the kth order position (i.e., the proximity between the adjacently-placed objects with subscripts j_{k-1} and j_k).

As usual, the validity of the recursive process requires the incremental cost index, $C((A_{k-1}, j_{k-1}), (A_k, j_k)) = p_{(j_{k-1})j_k}$, to depend only on $(A_{k-1}, j_{k-1}) \in \Omega_{k-1}$ and $(A_k, j_k) \in \Omega_k$, but in contrast to the GDPP specializations of Sections 4.1.1 to 4.1.3, we now know which was the last-placed subscript j_{k-1} in A_{k-1}; thus, cost increments can be defined using those objects with subscripts j_{k-1} and j_k as the last-placed indices in A_{k-1} and A_k, respectively. The values of $\mathcal{F}((A_1, j_1))$ can be assumed zero for all $(A_1, j_1) \in \Omega_1$, and the recursive process can be carried out from Ω_1 to Ω_n. The value defined by the minimum of $\mathcal{F}((A_n, j_n))$ over all j_n, $1 \leq j_n \leq n$, for $(A_n, j_n) \in \Omega_n$ and $A_n = \{1, 2, \ldots, n\}$ provides the optimal (minimal) value for the sum of adjacent proximities over all paths among the n objects in S. As always, an optimal row/column reordering of $\mathbf{P}$ attaining this minimal value can be identified by working backward through the recursion.

There are several variations on the optimal construction of a path that can be directly implemented through the type of GDPP specialization just described.

4.1. OPTIMAL SEQUENCING OF A SINGLE OBJECT SET

One is an interpretation through the use of merit increments (rather than cost increments), $M((A_{k-1}, j_{k-1}), (A_k, j_k)) = p_{(j_{k-1})j_k}$, and adopts a maximization optimization criterion. This change is immediate and offers no difficulty because the GDPP recursion given in (2.3) can be applied. Similarly, a min/max or max/min criterion can also be used merely by selecting the general recursive structure of (2.5) or (2.6), which would identify optimal paths among the objects in S that either minimize the maximum adjacent proximity or maximize the minimum adjacent proximity.[35] In contrast to the GDPP recursion of Sections 4.1.1 through 4.1.3, there are now greater storage requirements because in the construction of optimal paths, the sets $\Omega_k, 1 \leq k \leq n$, are defined by pairs $(A_k, j_k) \in \Omega_k$. In any case, this general type of recursive process defined on sets having this latter form was first described independently by Bellman (1962) and Held and Karp (1962) for what is called the traveling salesman problem. We will return to this specific topic briefly at the end of this section.

The symmetric proximity matrix $\mathbf{P}$ used in the construction of an optimal path between the objects in S has been considered arbitrary up to this point. There are, however, several substantive applications for this optimization task already suggested in the literature, which depend on specific definitions of how $\mathbf{P}$ may be constructed from other data available on the objects in S. To be explicit, and to mention briefly a few of these applications (for a more detailed review, see Hubert and Baker (1978)), suppose an $n \times p$ data matrix $\mathbf{X} = \{x_{ij}\}$ is given, where the rows of $\mathbf{X}$ refer to the objects in S, and the columns of $\mathbf{X}$ refer to p attributes measured on each of the n objects. Depending on how $\mathbf{P}$ is constructed, several data analysis applications can be given as specific exemplars:

Profile smoothing. Given an $n \times p$ data matrix $\mathbf{X}$, one visual means for displaying the information it contains is to first place the n objects along a horizontal axis and then graph the p profiles for each attribute over the n objects. Depending on the object order used along the horizontal axis (as discussed by Hartigan (1975), pp. 28-34, Späth (1980), Chapter 5, and Wegman (1990)), there may be a way of reducing the complexity of the graphical representation by minimizing the number of instances in which the profiles cross. If a proximity matrix $\mathbf{P} = \{p_{ij}\}$ is defined as

$$p_{ij} = \sum_{k<k'} g(x_{ik}, x_{ik'}, x_{jk}, x_{jk'}),$$

where

$$g(x_{ik}, x_{ik'}, x_{jk}, x_{jk'}) = \begin{cases} 1 & \text{if } (x_{ik} - x_{ik'})(x_{jk} - x_{jk'}) < 0; \\ 0 & \text{otherwise,} \end{cases}$$

then the ordering of the objects in S, minimizing the sum of proximities between adjacent objects along the path, also minimizes the number of instances in which the profiles cross.

Data array reordering. To help interpret the patterning of data present in $\mathbf{X}$, it may be of value to reorder the rows (and possibly the columns as well) of **X** so that the numerically larger elements of the array are placed as close as possible to each other. As discussed by McCormick, Schweitzer, and White (1972) (and clarified by Lenstra (1974); see Arabie and Hubert (1990) for a review), one possibility would be to define a proximity matrix $\mathbf{P} = \{p_{ij}\}$ among the n rows of **X** (now with a similarity interpretation) through a simple cross-product measure over the p column attributes, i.e.,

$$p_{ij} = \sum_k x_{ik} x_{jk}.$$

The optimal path sought among the n objects would maximize the sum of proximities between adjacent objects, and the object order thus obtained could be used to reorder the rows of **X**.

As a related suggestion for possibly reordering the rows of **X** to help interpret the pattern of information present, Kendall (1971a; 1971b) observed that if the rows of **X** could be reordered to display a perfect Q-form within columns (see Section 4.1.3), and if $\mathbf{P} = \{p_{ij}\}$ is defined as

$$p_{ij} = \sum_k \max(x_{ik}, x_{jk}),$$

then **P** can be reordered to display a perfect anti-Robinson form. Using this same reordering on the rows of **X**, a perfect Q-form within columns would be displayed. Thus, one possible strategy for attempting to find an approximate Q-form for **X** would be to identify the minimum length path using **P** and use the object order so identified to reorder the rows of **X**. (Such a method obviously depends on the result that if any proximity matrix **P** can be reordered to display a perfect anti-Robinson pattern, then the minimum-length path can be used to identify such an ordering.)

Numerical illustrations. As examples of constructing optimal paths based on a symmetric proximity matrix **P**, we again consider the before and after submatrices of Table 4.1 on the rated seriousness of thirteen offenses. The two optimal reorderings minimizing the sum of proximities between adjacently-placed objects are given graphically in Figure 4.1, where the respective optimal lengths are 2.30 and 2.34 for the before and after data. The two orders are very consistent with the results given earlier (e.g., with the explicit coordinate representation using the measure of matrix pattern in (4.6)); again, there are some differences among offenses 5, 6, and 13, which are very close to one another.

Although our discussion of constructing optimal paths based on **P** has been phrased as obtaining a sequence of adjacent objects that includes each object in S exactly once and some function of the $n - 1$ proximities between adjacently placed objects, the more traditional discussion of optimal path construction in the literature (e.g., see Lawler et al. (1985) for an extensive review in book form) is concerned with the generation of optimal circular paths in which each object is also included exactly once but the path is closed and now includes n proximities between the adjacently-placed objects. This topic is typically discussed

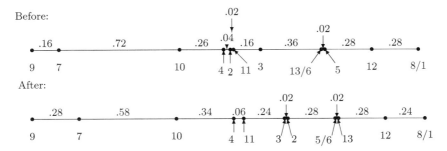

Figure 4.1: *The optimal orders for the before and after matrices of Table 4.1 minimizing the sum of proximities between adjacently-placed objects.*

under the label of the Traveling Salesman Problem, where interpretatively, a salesperson must visit each of n cities once and only once, return to the city of origin, and minimize the length of the tour. The type of GDPP recursion used for the construction of optimal linear paths can be modified easily for the construction of optimal circular paths:

Choose object O_1 as an (arbitrary) origin and force the construction of the optimal linear paths to include O_1 as the initial object by defining $\mathcal{F}((A_1, j_1)) = 0$ for $j_1 = 1$ and $A_1 = \{1\}$, and otherwise equal to a very large positive or negative value (depending on whether the task is a minimization or a maximization, respectively). The function values $\mathcal{F}((A_n, j_n))$ for all $j_n, 1 \leq j_n \leq n$, for $(A_n, j_n) \in \Omega_n$, and for $A_n = \{1, 2, \ldots, n\}$ can then be used to obtain the optimal circular paths depending on the chosen optimization criteria as follows:

minimum path length: $\min[\mathcal{F}((A_n, j_n)) + p_{j_n 1}]$;

maximum path length: $\max[\mathcal{F}((A_n, j_n)) + p_{j_n 1}]$;

minimax path length: $\min[\max(\mathcal{F}((A_n, j_n)), p_{j_n 1})]$;

maximin path length: $\max[\min(\mathcal{F}((A_n, j_n)), p_{j_n 1})]$.

Numerical illustrations (continued). As an example of how such an optimization might be carried out, Figure 4.2 represents an optimal circular path for the proximity matrix from Shepard, Kilpatric, and Cunningham (1975) on the ten digits of Table 1.1, minimizing the sum of the ten pairwise input proximities between the adjacently-placed objects (the minimal value is 3.411). Interpretatively, the multiples of 2 (2,4,8) and of 3 (3,6,9) appear at adjacent locations, with the odd numbers that are not multiples of 3 (5,7) and the identities (0,1) placed between these two groups and arranged to be as consistent as possible with digit magnitude, e.g., 5 and 6, and 7 and 8 are adjacent; and 0 and 1 are placed close to 2 and 3.

In closing this section, we make three final observations about the type of GDPP recursive process discussed above. First, although it has been explicitly assumed that the proximity matrix $\mathbf{P}$ is symmetric, if an $n \times n$ nonsymmetric

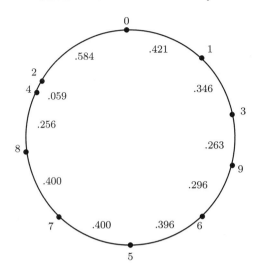

Figure 4.2: *An optimal circular ordering of the ten digits that minimizes the sum of proximities from Table 1.1 between adjacently-placed objects.*

matrix were to be used, then the same recursive process could be implemented to construct optimal *directed* linear or circular paths among the n objects in S, where each link in the path has an implied direction, i.e., p_{ij} (p_{ji}) is the proximity from object O_i to O_j (or from object O_j to O_i).[36] Second, it would be possible to redefine further the basic sets $\Omega_k, 1 \leq k \leq n$, to include additional specific information about the placement of certain objects other than just the last, e.g., we might define Ω_k to be (A_k, j_k, l_k), where j_k is the last-placed object in the set A_k and l_k denotes the second to the last. Although possible, this latter type of extension would obviously require much greater storage space; therefore, we will not explicitly pursue any of these generalizations here. Finally, although we have suggested the use of a GDPP recursive solution for the task of locating optimal paths (given the obvious emphases of this monograph), there are alternative optimization approaches that would be much better for dealing with large object sets and which could also provide optimal solutions (much as for the LA task introduced in Section 2.1). A detailed and comprehensive survey of these optimization options for the construction of optimal paths is available in the previously cited volume edited by Lawler et al. (1985).

4.2 Sequencing an Object Set Subject to Precedence Constraints

In the presentation of object sequencing in the various sections of 4.1, the search for optimality using the GDPP was carried out without any constraints on the

4.2. SEQUENCING SUBJECT TO PRECEDENCE CONSTRAINTS

object orderings. In all the cases discussed, however, it is straightforward to include precedence constraints that must be satisfied by the object orderings through the inclusion of an $n \times n$ indicator matrix $\mathbf{V} = \{v_{ij}\}$. The latter is defined over object pairs formed from S (where S may be $S_A \cup S_B$ for a two-mode proximity matrix), by letting $v_{ij} = 1$ if O_j must precede O_i in an (optimal) ordering, and 0 otherwise. Thus, by the device of incorporating $\mathbf{V}$ and not allowing transitions to occur between entities in Ω_{k-1} and Ω_k whenever the precedence conditions are not met, an optimal sequencing of S (or a joint sequencing of S_A and S_B) is sought that must satisfy all the constraints formalized by $\mathbf{V}$.[37]

Although precedence constraints can be incorporated directly into the use of the GDPP by the mechanism of including a precedence matrix $\mathbf{V}$, it is also possible, at least for a two-mode proximity matrix and in the joint sequencing of S_A and S_B, to use precedence constraints given by linear orderings (for example) of the row and/or column objects to reduce the storage requirements needed by an application of the GDPP. In particular, if the row and/or column objects in S_A and S_B are subject to linear order constraints, the sets $\Omega_1, \ldots, \Omega_K$ over which the recursive process is carried out can be redefined, thereby allowing larger object sets to be jointly sequenced optimally.

To be explicit, if there is an assumed linear ordering of the column objects in $S_B = \{c_1, \ldots, c_{n_B}\}$ (but none is assumed for the row objects in $S_A = \{r_1, \ldots, r_{n_A}\}$) that without loss of generality can be taken in the subscript order $c_1 \prec c_2 \prec \cdots \prec c_{n_B}$, then Ω_k for $1 \leq k \leq n$ may be defined so that each entity in Ω_k is of the form (B_k, j_k), where j_k is an integer from 0 to k indicating that the first j_k column objects have been placed, and B_k is a $k - j_k$ element subset of the row object subscripts $\{1, \ldots, n_A\}$ (which is the empty subset $\emptyset$ when $k - j_k = 0$). The function value $\mathcal{F}((B_k, j_k))$ denotes the optimal contribution to the total measure of matrix patterning when the first k positions in the reordering are occupied by the first j_k column objects and the $k - j_k$ row objects in B_k. Thus, because Ω_1 contains the members $(\emptyset, 1), (\{1\}, 0), \ldots, (\{n_A\}, 0)$, and assuming, say, the use of the weighted or unweighted gradient measures of matrix pattern of Section 4.1.3, $\mathcal{F}((B_1, j_1))$ can be assumed zero for all $(B_1, j_1) \in \Omega_1$, and the recursive process carried out from Ω_1 to Ω_n. The possible transformations between $(B_{k-1}, j_{k-1}) \in \Omega_{k-1}$ and $(B_k, j_k) \in \Omega_k$ are

(1) $j_{k-1} = j_k$ and $B_{k-1} \subset B_k$, where $B_{k-1} - B_k$ contains a single row object placed at the kth order position;

(2) $j_{k-1} + 1 = j_k$ and $B_{k-1} = B_k$, where the single column object with subscript j_k is placed at the kth order position.

The optimal value is achieved for the one entity $(\{1, \ldots, n_A\}, n_B) \in \Omega_n$, i.e., $\mathcal{F}((\{1, \ldots, n_A\}, n_B))$; as usual, working backward through the recursion identifies an optimal ordering of the $n = n_A + n_B$ objects in $S = S_A \cup S_B$, where the column objects in S_B appear in the subscript order $c_1 \prec c_2 \prec \cdots \prec c_{n_B}$. (Obviously, order constraints on the row objects alone [and not on the column objects] could be handled merely by reversing the roles of the row and column objects in the original $n_A \times n_B$ proximity matrix $\mathbf{Q}$.)

If linear orderings can be assumed for *both* the row and column objects in S_A and S_B (that again, without loss of generality, can be taken in the subscript orders $c_1 \prec c_2 \prec \cdots \prec c_{n_B}$ and $r_1 \prec r_2 \prec \cdots \prec r_{n_A}$, respectively), an even more substantial reduction in storage requirements can be achieved. Here, the sets Ω_k for $1 \leq k \leq n$ would be defined so that each entity in Ω_k is of the form (i_k, j_k), where both i_k and j_k are integers within the range 0 to k, subject to $i_k + j_k = k$, which indicates that the first i_k row objects and the first j_k column objects have been placed. The function value $\mathcal{F}((i_k, j_k))$ denotes the optimal contribution to the total measure of matrix patterning when the first k positions in the reordering are occupied by the first i_k and the first j_k row and column objects. The possible transformations between $(i_{k-1}, j_{k-1}) \in \Omega_{k-1}$ and $(i_k, j_k) \in \Omega_k$ are

(1) $j_k = j_{k-1}$ and $i_k = i_{k-1} + 1$;
(2) $j_k = j_{k-1} + 1$ and $i_k = i_{k-1}$.

Beginning with Ω_1, which contains the entities (i_1, j_1), where $i_1 + j_1 = 1$ (i.e., the two pairs $(0,1)$ and $(1,0)$) and $\mathcal{F}((i_1, j_1)) = 0$ (for, say, the weighted or unweighted gradient measures of matrix pattern of Section 4.1.3), the recursive process proceeds from Ω_1 to Ω_n. The latter contains the single entity (n_A, n_B), and $\mathcal{F}((n_A, n_B))$ defines the optimal value for the chosen measure of matrix pattern. The optimal ordering is again identified by working backward through the recursion from Ω_n to Ω_1.

The type of DP recursive process just described for finding a joint sequencing of the sets S_A and S_B when both are subject to linear order constraints was first described in Delcoigne and Hansen (1975) for a particular measure of matrix patterning discussed by Gordon (1973). This measure of matrix patterning (to be referred to as the DHG measure) for a joint sequencing of row and column objects is the sum of the proximities between the row objects and the adjacently-placed column objects plus the proximities between the column objects and the adjacently-placed row objects. So, assuming proximity has a dissimilarity interpretation, an optimal joint sequencing would minimize the DHG measure. (As one technicality, if there is only a single adjacent column object for a particular row object in the joint sequence, then the proximity to this single adjacent column object is doubled; similarly, if there is only a single adjacent row object for a particular column object in the joint sequence, then the proximity to this single adjacent row object is doubled). The same form of the recursive process just described can be carried out for the DHG measure by defining $\mathcal{F}((i_1, j_1))$ for $(i_1, j_1) \in \Omega_1$ to be $2q_{11}$ and the cost increment in moving from $(i_{k-1}, j_{k-1}) \in \Omega_{k-1}$ to $(i_k, j_k) \in \Omega_k$ to be either

(1) $q_{i_k j_k} + q_{i_k (j_k+1)}$ when $j_k = j_{k-1}$ and $i_k = i_{k-1}+1$ (if $j_k = 0$, the increment is $2q_{i_k(j_k+1)}$; if $j_k = n_B$, the increment is $2q_{i_k j_k}$);

(2) $q_{i_k j_k} + q_{(i_k+1) j_k}$ when $j_k = j_{k-1}+1$ and $i_{k-1} = i_k$ (if $i_k = 0$, the increment is $2q_{(i_k+1) j_k}$; if $i_k = n_A$, the increment is $2q_{i_k j_k}$).[38]

4.2. SEQUENCING SUBJECT TO PRECEDENCE CONSTRAINTS

Numerical illustrations. To provide an example of imposing row and/or column constraints, we will again use the 11×9 data matrix of Table 1.3 on the dissimilarities between the eleven goldfish retinal receptors and the nine specific wavelengths of light. In Section 4.1 the optimal orderings were provided for the combined row/column object set based on both the unweighted and weighted gradient measures. The row orders for the two gradient measures differed by an adjacent receptor interchange for 8 and 9; the separate column orders differed by the adjacent wavelength interchange for $\underline{4}$ and $\underline{9}$ with the unweighted gradient measure giving an ordering that was completely consistent with decreasing wavelengths. To illustrate the effect of imposing row and/or column constraints, we use the weighted gradient measure in the comparisons below but impose for the column constrained joint sequencing that column order obtained for the unweighted gradient measure (i.e., $\underline{3} \to \underline{8} \to \underline{2} \to \underline{7} \to \underline{1} \to \underline{5} \to \underline{6} \to \underline{4} \to \underline{9}$). When the additional row constraints are to be imposed, we use the row order obtained also from the unweighted gradient measure (i.e., $10 \to 11 \to 5 \to 7 \to 1 \to 4 \to 6 \to 2 \to 8 \to 9 \to 3$). To allow comparison, the unrestricted joint sequencing of S_A and S_B from Section 4.1 is also reproduced below based on the weighted gradient measure; also, in all cases the descriptive ratios obtained from (4.4) are provided.

Unrestricted joint sequencing of S_A and S_B, obtained using DPSE2U (taken from Section 4.1):

$\underline{3}$ $\underline{8}$ 10 11 $\underline{2}$ 5 7 1 $\underline{7}$ $\underline{1}$ 4 6 2 $\underline{5}$ $\underline{6}$ $\underline{9}$ 9 $\underline{4}$ 8 3

weighted gradient measure $= 31025 = (32019 - 994)$
ratio $= .940 = (32019 - 994)/(32019 + 994)$

Column order restricted joint sequencing of S_A and S_B, obtained using DPSE2R:

$\underline{3}$ $\underline{8}$ 10 11 $\underline{2}$ 5 7 1 $\underline{7}$ $\underline{1}$ 4 6 2 $\underline{5}$ $\underline{6}$ [9 $\underline{4}$ $\underline{9}$] 8 3

weighted gradient measure $= 30985 = (31998 - 1013)$
ratio $= .939 = (31998 - 1013)/(31998 + 1013)$

Row and column order restricted joint sequencing of S_A and S_B, obtained using DPSE2R:

$\underline{3}$ $\underline{8}$ 10 11 $\underline{2}$ 5 7 1 $\underline{7}$ $\underline{1}$ 4 6 2 $\underline{5}$ $\underline{6}$ [$\underline{4}$ $\underline{9}$ 8 9] 3

weighted gradient measure $= 30867 = (31933 - 1066)$
ratio $= .935 = (31933 - 1066)/(31933 + 1066)$

DHG index of matrix patterning:

$\underline{3}$ 10 $\underline{8}$ 11 5 7 $\underline{2}$ 1 4 $\underline{7}$ 6 $\underline{1}$ 2 $\underline{5}$ $\underline{6}$ 8 $\underline{4}$ 9 3 $\underline{8}$

DHG index $= 2317.0$

In comparison to the unrestricted joint sequencing of S_A and S_B, the imposition of row and/or column constraints does have an effect, albeit somewhat small, in the optimal weighted gradient measures. Also, the comparison of the joint sequencings for the DHG and weighted gradient measures does suggest that the specific type of intermixing of the row and column objects will be influenced heavily by the choice of index.

Although we will not pursue the topic in any further detail here, it may be of some interest to note an extensively developed area in the literature that addresses the task of comparing two linearly-ordered object sets (in contrast to combining them and finding an optimal joint sequencing), where the preferred strategy of comparison is again through a DP recursive process. The volume edited by Sankoff and Kruskal (1983) provides a variety of applications to the comparison of genetic sequences, time-warping problems in the processing of speech, and string-correction methods in computer science, among others. In our notation, the comparison process can be characterized as follows: we are given two sets $S_A = \{r_1, \ldots, r_{n_A}\}$ and $S_B = \{c_1, \ldots, c_{n_B}\}$, where it is assumed that the linear orderings within S_A and S_B are $r_1 \prec r_2 \prec \cdots \prec r_{n_A}$ and $c_1 \prec c_2 \prec \cdots \prec c_{n_B}$. Each of these sequences can be augmented by the inclusion of null elements, say, $\oslash$, to produce two new sequences, $r'_1 \prec r'_2 \prec \cdots \prec r'_{n_A+n_B}$ and $c'_1 \prec c'_2 \prec \cdots \prec c'_{n_A+n_B}$, where each r'_i is either a null element $\oslash$ or r_j for some integer j, $1 \leq j \leq n_A$, and each c'_i is either a null element $\oslash$ or c_j for some integer j, $1 \leq j \leq n_B$, and the nonnull entities in either sequence are in their given linear ordering. The measure of comparison between the two sequences is based on the one-to-one correspondence between r'_i and c'_i, i.e.,

$$\sum_{i=1}^{n_A+n_B} u(r'_i, c'_i), \qquad (4.9)$$

where the costs $u(r'_i, c'_i)$ are assumed to be given and depend on whether r'_i and c'_i are both nonnull entities, or whether either r'_i or c'_i or both are null entities (it can be supposed that $u(r'_i, c'_i) = 0$ whenever r'_i and c'_i are both null entities because they could merely be deleted from their respective sequences and, thus, would play no role in assessing the dissimilarity of the two sequences). Computationally, we wish to minimize the index in (4.9) and find an optimal matching between the two sequences (as augmented by the possible inclusion of null entities in each); the optimal value is assumed to provide a reasonable measure of dissimilarity between the two original (linearly-ordered) sequences.

An application of the general form of the GDPP to minimize the index in (4.9) is very direct. The sets $\Omega_1, \ldots, \Omega_n$ (where $n = n_A + n_B$) can be defined so that Ω_k contains the ordered pairs (i_k, j_k), where i_k and j_k are integers ($0 \leq i_k \leq k$; $0 \leq j_k \leq k$) and refer to the indices in the ordered sets S_A and S_B matched up to this point, i.e., i_k row objects and $k - i_k$ null entities have been matched with j_k column objects and $k - j_k$ null entries. The set Ω_1 contains four members, $(0, 0), (0, 1), (1, 0)$, and $(1, 1)$, that correspond to the matching that occurs in the first position, i.e., this matching can be, respectively, two null entities, a null entity and c_1, r_1 and a null entity, or r_1 and c_1. The values $\mathcal{F}((i_1, j_1))$ for $(i_1, j_1) \in \Omega_1$ are assumed given as $u(\oslash, \oslash)$, $u(\oslash, c_1)$, $u(r_1, \oslash)$, and $u(r_1, c_1)$ for these four pairs. A transformation of an entity in $(i_{k-1}, j_{k-1}) \in \Omega_{k-1}$ to an entity $(i_k, j_k) \in \Omega_k$ is possible if

(a) $i_{k-1} = i_k, j_{k-1} + 1 = j_k$ (with a cost increment of $u(\oslash, c_{j_k})$);
(b) $i_{k-1} + 1 = i_k, j_{k-1} = j_k$ (with a cost increment of $u(r_{i_k}, \oslash)$);
(c) $i_{k-1} + 1 = i_k, j_{k-1} + 1 = j_k$ (with a cost increment of $u(r_{i_k}, c_{j_k})$);
(d) $i_{k-1} = i_k, j_{k-1} = j_k$ (with a cost increment of $u(\oslash, \oslash) \equiv 0$).

Using the minimization form of the GDPP in (2.4), the recursion proceeds from Ω_1 to Ω_n, where Ω_n contains the single entity (n_A, n_B). The optimal value for a matching of the two ordered sequences is defined by $\mathcal{F}((n_A, n_B))$, with the actual solution (or matching) obtained by working backward through the recursive process. As noted, the very extensive literature on this comparison task can be accessed through the volume edited by Sankoff and Kruskal (1983).

4.3 Construction of Optimal Ordered Partitions

In Section 3.1, which dealt with the partitioning of an object set, the classes of an optimal partition, based on some measure of subset heterogeneity, had no particular order imposed on them. In contrast, the emphasis in Chapter 4, up to this point, has been on the optimal reordering of an object set, but this task has effectively been carried out by placing one object at a time. As a possible (and obvious) conjunction of these two tasks of partitioning and sequencing, this section discusses the problem of constructing an optimal (ordered) partition of an object set in which the classes of the partition must be sequenced along the continuum. We will only consider one-mode proximity data that come in the usual form of a symmetric matrix $\mathbf{P}$ or a skew-symmetric matrix $\mathbf{P}^{SS}$, and the relevant measures of matrix patterning from Sections 4.1.1 and 4.1.2 will be generalized. There are, as might be expected, a host of variations and extensions that could be pursued, which would parallel many topics discussed earlier in this monograph. Some of these possibilities will at least be noted at the end of this section.[39]

The basic task of constructing an ordered partition of an object set $S = \{O_1, \ldots, O_n\}$ into M ordered classes, $S_1 \prec S_2 \prec \cdots \prec S_M$, using some (merit) measure of matrix patterning and a proximity matrix $\mathbf{P}$ or $\mathbf{P}^{SS}$, can be approached through the general type of GDPP recursive process applied for the partitioning task of Section 3.1 but with appropriate variation in defining the merit increments. Explicitly, the sets $\Omega_1, \ldots, \Omega_K$ (where $K \equiv M$) will each contain all $2^n - 1$ nonempty subsets of the n object subscripts; $\mathcal{F}(A_k)$ for $A_k \in \Omega_k$ is the optimal value for placing k classes in the first k positions, and the subset A_k is the union of these k classes. A transformation from $A_{k-1} \in \Omega_{k-1}$ to

$A_k \in \Omega_k$ is possible if $A_{k-1} \subset A_k$; the merit increment $M(A_{k-1}, A_k)$ is based on placing the class $A_{k-1} - A_k$ at the kth position (which will depend on A_{k-1}, A_k, and $S - A_k$). Beginning with $\mathcal{F}(A_1)$ for all $A_1 \in \Omega_1$ (i.e., the merit of placing the class A_1 at the first position), the recursion proceeds from Ω_1 to Ω_K, with $\mathcal{F}(A_K)$ for $A_K = S \in \Omega_K$ defining the optimal merit value for an ordered partition into $K (\equiv M)$ classes (which can then be identified as usual by working backward through the recursion). Also, optimal ordered partitions that contain from 2 to $M - 1$ classes are identified by $\mathcal{F}(A_2), \ldots, \mathcal{F}(A_{M-1})$ when $S = A_2 = \cdots = A_{M-1}$. It is necessary to specify in a particular context the merit increment, $M(A_{k-1}, A_k)$; this is discussed below for various measures of matrix patterning generalized from Sections 4.1.1 and 4.1.2.

For a Symmetric Proximity Matrix P:

To generalize the weighted or unweighted gradient measure given in (4.1), the merit increment for placing the *class* $A_k - A_{k-1}$ at the kth order position is $I_{row}(A_k - A_{k-1}) + I_{col}(A_k - A_{k-1})$, where

$$I_{row}(A_k - A_{k-1}) = \sum_{i' \in A_{k-1}} \sum_{k' \in A_k - A_{k-1}} \sum_{j' \in S - A_k} f(p_{i'k'}, p_{i'j'})$$

and (4.10)

$$I_{col}(A_k - A_{k-1}) = \sum_{i' \in A_{k-1}} \sum_{k' \in A_k - A_{k-1}} \sum_{j' \in S - A_k} f(p_{k'j'}, p_{i'j'}).$$

To initialize the recursion, we let $\mathcal{F}(A_1) = 0$ for all $A_1 \in \Omega_1$.

A merit measure based on a coordinate representation for each of the M ordered classes, $S_1 \prec S_2 \prec \cdots \prec S_M$, that generalizes (4.6) can also be developed directly. Here, M coordinates, $x_1 \leq \cdots \leq x_M$, are to be identified so that the residual sum-of-squares,

$$\sum_{k \leq k'} \sum_{i_k \in S_k,\ j_{k'} \in S_{k'}} (p_{i_k j_{k'}} - \mid x_{k'} - x_k \mid)^2,$$

is minimized (the notation $p_{i_k j_{k'}}$ indicates those proximities in **P** defined between objects with subscripts $i_k \in S_k$ and $j_{k'} \in S_{k'}$). A direct extension of the argument that led to optimal coordinate representation for single objects would require the maximization of

$$\sum_{k=1}^{M} \left(\frac{1}{n_k}\right) (G(A_k - A_{k-1}))^2, \qquad (4.11)$$

where $G(A_k - A_{k-1}) =$

$$\sum_{k' \in A_k - A_{k-1}} \sum_{i' \in A_{k-1}} p_{k'i'} - \sum_{k' \in A_k - A_{k-1}} \sum_{i' \in S - A_k} p_{k'i'},$$

4.3. CONSTRUCTION OF OPTIMAL ORDERED PARTITIONS

and n_k denotes the number of objects in $A_k - A_{k-1}$. The merit increment for placing the subset $A_k - A_{k-1}$ at the kth order position would be $(1/n_k)(G(A_k - A_{k-1}))^2$, with the recursion initialized by

$$\mathcal{F}(A_1) = \left(\frac{1}{n_1}\right)\left(\sum_{k' \in A_1} \sum_{i' \in S - A_1} p_{k'i'}\right)^2$$

for all $A_1 \in \Omega_1$. If an optimal ordered partition that maximizes (4.11) is denoted by $S_1^* \prec \cdots \prec S_M^*$, the optimal coordinates for each of the M classes can be given as

$$x_k^* = \left(\frac{1}{nn_k}\right) G(S_k^*),$$

where $x_1^* \leq \cdots \leq x_M^*$ and $\sum_k n_k x_k^* = 0$. The residual sum-of-squares has the form

$$\sum_{i<j} p_{ij}^2 - \left(\frac{1}{n}\right) \sum_k \left(\frac{1}{n_k}\right)(G(S_k^*))^2.$$

For a Skew-Symmetric Proximity Matrix $\mathbf{P}^{SS}$:

In extending the measure in (4.7) for skew-symmetric matrices that led to maximizing the above-diagonal entries in $\mathbf{P}_\rho^{SS}$, the merit increment is now defined as

$$J(A_k - A_{k-1}) = \sum_{i' \in A_{k-1}} \sum_{k' \in A_k - A_{k-1}} p_{i'k'},$$

to indicate the contribution from the class $A_k - A_{k-1}$ when placed at the kth order position. Again, the recursive process is initiated by letting $\mathcal{F}(A_1) = 0$ for all $A_1 \in \Omega_1$.

The definition of a merit measure based on a coordinate representation for a skew-symmetric matrix as discussed in Section 4.1.2 is very direct when each object is treated separately because a closed-form solution was possible for the optimal coordinates. The situation changes, however, when attempting to obtain an optimally ordered partition based on $\mathbf{P}^{SS}$ through a coordinate representation. Explicitly, suppose we want to find a set of M coordinates, $x_1, \ldots, x_M$, and a set of M classes, $S_1, \ldots, S_M$ (not necessarily ordered), such that

$$\left(\frac{1}{2}\right) \sum_{k,k'} \sum_{i_k \in S_k,\ j_{k'} \in S_{k'}} (p_{i_k j_{k'}}^{SS} - (x_k - x_{k'}))^2 \quad (4.12)$$

is minimized, where again $p_{i_k j_{k'}}^{SS}$ indicates a proximity between objects with subscripts $i_k \in S_k$ to $j_{k'} \in S_{k'}$. If the classes $S_1, \ldots, S_M$ with respective sizes $n_1, \ldots, n_M$ were given, the M coordinates, say, $x_1, \ldots, x_M$, could be obtained by

$$x_k = \left(\frac{1}{nn_k}\right) \sum_{k' \in S_k} \sum_{i' \in S - S_k} p_{k'i'}^{SS}.$$

There is, however, a preliminary need to choose $S_1, \ldots, S_M$ optimally if the loss function in (4.12) is to be minimized. This optimal choice can be carried out by defining an increment of merit in the GDPP for placing the subset $A_k - A_{k-1}$ at the kth order position, as $(1/n_k)(L(A_k - A_{k-1}))^2$, where

$$L(A_k - A_{k-1}) = \sum_{k' \in A_k - A_{k-1}} \sum_{i' \in S - (A_k - A_{k-1})} p_{k'i'}^{SS},$$

and initializing the recursion by

$$\mathcal{F}(A_1) = \left(\frac{1}{n_1}\right) \left(\sum_{k' \in A_1} \sum_{i' \in S - A_1} p_{k'i'}^{SS} \right)^2.$$

The optimal classes thus identified, say, $S_1^*, \ldots, S_M^*$, lead directly to the M optimal coordinates $x_1^*, \ldots, x_M^*$, where

$$x_k^* = (1/nn_k) \sum_{k' \in S_k^*} \sum_{i' \in S - S_k^*} p_{k'i'}^{SS},$$

and a residual sum-of-squares of

$$\sum_{i<j} (p_{ij}^{SS})^2 - n \sum_{k=1}^{n} n_k (x_k^*)^2.$$

Numerical illustrations.[40] For constructing optimal ordered partitions, we will again consider the data of Table 4.1 on the rated seriousness of thirteen offenses both before and after viewing a film, and also both its skew-symmetric and symmetric forms (where the latter are based on taking absolute values of the skew-symmetric proximities). For brevity, we will present for the various measures of matrix patterning the original ordering using single objects (i.e., partitions with thirteen ordered classes) and an optimally ordered partition with five classes. In all cases, there was a very precipitous change in the merit measures when moving from five to four classes, and, therefore, the choice of presenting only those optimal ordered partitions with five classes is not arbitrary. As shown in the results summarized below, the five ordered (according to increasing severity) classes of offenses consistently include 'gambler' (#2) in the second class before viewing the film and in the third class after (we might also comment that in most of the analyses reported, the 5-class and the 13-class ordered partitions are completely consistent in the sense that the classes in the former are defined by consecutively-placed single objects in the latter; the exceptions are for the gradient measures and a few adjacently-located objects). In general, the ordered partitions into thirteen and five classes are very similar within the before and within the after conditions, and are consistent over the various measures of matrix patterning for either the symmetric or skew-symmetric proximity matrices (although for completeness we give the exhaustive listing of results for all the options below).

4.3. CONSTRUCTION OF OPTIMAL ORDERED PARTITIONS

Optimal ordered partitions into thirteen and five classes, obtained from the program DPOP1U:

Before viewing:
symmetric proximity; coordinate representation

object order	9	7	10	4	2	11	3	5	13	6	12	8	1
coordinates	−.82	−.78	−.33	−.26	−.23	−.17	−.02	.27	.29	.32	.50	.59	.64
(5 classes):	{	−.80	}	{	−.25		}	{−.02}	{	.29	}	{	.58 }

residual sum-of-squares: (13 classes) 3.307; (5 classes) 3.635

symmetric proximity; unweighted gradient
object order 9 7 4 10 2 11 3 5 13 6 12 8 1
(5 classes) { } { } { } {13 6 8} {12 1}
index of gradient comparisons: (13 classes) 431; (5 classes) 289

symmetric proximity; weighted gradient
object order 9 7 10 4 2 11 3 5 13 6 12 8 1
(5 classes) { } { } { } {6 8} {12 1}
index of gradient comparisons: (13 classes) 161.45; (5 classes) 100.08

skew-symmetric proximity; above-diagonal sum
object order 9 7 10 2 4 11 3 6 5 13 1 12 8
(5 classes) { } { } { } { } { }
above-diagonal sum: (13 classes) 49.93; (5 classes) 48.27

skew-symmetric proximity; coordinate representation

object order	9	7	10	4	2	11	3	5	13	6	1	12	8
coordinates	−.82	−.78	−.33	−.26	−.23	−.18	−.02	.27	.29	.32	.57	.57	.59
(5 classes)	{	−.80	}	{	−.25		}	{−.02}	{	.29	}	{	.58 }

residual sum-of-squares: (13 classes) 3.457; (5 classes) 3.635

After viewing:
symmetric proximity; coordinate representation

object order	9	7	10	4	11	3	2	6	13	5	12	8	1
coordinates	−.81	−.75	−.39	−.26	−.21	−.05	−.02	.27	.27	.29	.48	.55	.58
(5 classes)	{	−.78	}	{	−.29		}	{−.02}	{	.28	}	{	.54 }

residual sum-of-squares: (13 classes) 2.302; (5 classes) 2.674

symmetric proximity; unweighted gradient
object order 9 7 10 4 11 3 2 13 6 5 12 8 1
(5 classes) { } { } { } { } { }
index of gradient comparisons: (13 classes) 491; (5 classes) 331

symmetric proximity; weighted gradient
object order 9 7 10 4 11 3 2 6 13 5 12 8 1
(5 classes) {9 7 4} {10 11 3} { } {5 8} {12 1}
index of gradient comparisons: (13 classes) 165.08; (5 classes) 101.84

skew-symmetric proximity; above-diagonal sum
object order 9 7 10 4 11 3 2 5 6 13 1 12 8
(5 classes) { } { } { } { } { }
above-diagonal sum: (13 classes) 47.42; (5 classes) 45.86

skew-symmetric proximity; coordinate representation

object order	9	7	10	4	11	3	2	6	5	13	12	1	8
coordinates	−.81	−.75	−.39	−.26	−.21	−.05	.02	.27	.28	.29	.51	.55	.55
(5 classes)	{	−.78	}	{	−.29		}	{ −.02	}	{	.28	}	{ .54 }

residual sum-of-squares: (13 classes) 2.369; (5 classes) 2.674

The task of identifying optimal ordered partitions has been limited to a discussion of one-mode matrices, but there are numerous variations that could be pursued, related to topics already raised in earlier sections. For example, extensions would be possible to the use of order (or precedence) restrictions on a one-mode matrix (as in Section 3.1.1), or to two-mode proximity matrices that may be row and/or column order (or precedence) restricted (as in Section 3.1.2). Similarly, the classes of the ordered partitions as M varies from 1 to n might be restricted to be hierarchical (as in Section 3.2) and with or without a consecutive order restriction on which objects can form the classes of each partition. Relatedly, as in the discussion of object sequencing in Sections 4.1 and 4.2, gradient comparisons might be restricted to be only within rows or within columns of a one-mode proximity matrix or, for a two-mode matrix, only within the rows or columns. Alternative gradient measures could also be adopted, e.g., minimizing only the gradient violations, attempting to use the Greenberg form, or concentrating on an equally spaced coordinate representation. For a further discussion of the task of constructing optimal ordered partitions, the reader is referred to Hubert, Arabie, and Meulman (1997b).

Endnotes

[27] Because the general recursive process just described requires the storage of intermediate results for all possible subsets of an object set with n members, the two programs to be used in the next three sections, DPSE1U and DPSE2U (where the suffix '1U' refers to '1-mode unrestricted' and '2U' to '2-mode unrestricted'), are effectively limited to object set sizes in their low 20's given the typical RAM configurations currently available, although no formal upper limits are built into either program.

[28] Although we choose not to do so, the discussion of the GDPP applications in this chapter could employ the terminology of posets (see Chapter 2, endnote 6), as well as several more restrictive concepts usually introduced in that framework, e.g., lattices, maximal chains, and the like. To be a little more specific (but still only in a very schematic form based on the notation in Chapter 2, endnote 6), the set Ω would contain all partitions of S, and the relation, $\preceq$, would be defined by partition refinement $A \preceq A'$ (in other words, A is a "refinement" of A') if all the classes in A are in A' or can be formed from subdividing those present in A'. The pair $(\Omega, \preceq)$ is a poset; oreover, it is a lattice in that any two elements A and A' in Ω have a greatest-lower-bound, denoted $A \wedge A'$ (and read as A "meet" A' or as the "meet" of A and A'), and a least-upper-bound, denoted $A \vee A'$ (and read as A "join" A' or as the "join" of A and A'). Formally, $A \wedge A'$ is the unique element of Ω such that $A \wedge A' \preceq A$, $A \wedge A' \preceq A'$, and for any other member A'' of Ω, if $A'' \preceq A$ and $A'' \preceq A'$, then $A'' \preceq A \wedge A'$. Constructively, $A \wedge A'$ is generated from all pairwise intersections of the classes in A and A'. The partition $A \vee A'$ is the unique element of Ω such that $A \preceq A \vee A'$, $A' \preceq A \vee A'$, and for any other member A'' of Ω, if $A \preceq A''$ and $A' \preceq A''$, then $A \vee A' \preceq A''$. Constructively, $A \vee A'$ is generated from the meet of all partitions in Ω that are greater than or equal to both A and A' according to the order relation $\preceq$.

The set Ω is a lattice under the join and meet operations, and any nonempty subset of Ω, say, Ω', that is also a lattice under the same operations is called a

4.3. CONSTRUCTION OF OPTIMAL ORDERED PARTITIONS 89

sublattice of Ω. If, in addition, A and A' in Ω' imply $A \preceq A'$ or $A' \preceq A$, then Ω' is said to be totally (or simply) ordered and is called a chain. Consequently, the later discussions in Section 3.2 of constructing a hierarchical sequence of partitions could be phrased, if we wished, as finding totally ordered sublattices or chains. Also, in our later use of the term "full partition hierarchy," where successive partitions are constructed by uniting only a single pair of classes from the one given previously, we could refer to a maximal chain, i.e., a totally ordered sublattice of Ω in which each element except the first covers its predecessor.

[29] There is now a rather extensive literature on graphically representing a matrix having either a Robinson or an anti-Robinson form (see Chapter 3, endnote 13, for the distinction between these two terms). In this monograph our emphasis is solely on the main combinatorial optimization tasks, and in this chapter specifically on identifying optimal object orders for a proximity matrix; therefore, we will not go further into the subsidiary issues of graphically representing an (anti-) Robinson matrix that is being used as an approximation. The reader interested in pursuing some of the relevant literature might begin with Diday (1986) and the introduction to graphically representing an (anti-)Robinson matrix by pyramids, and then continue on with the review by Durand and Fichet (1988), who point out the necessity of strengthening the basic (anti-)Robinson condition to one that is strongly-(anti-)Robinson if a consistent graphical (pyramidal) representation is to be possible — otherwise, unresolvable graphical anomalies can arise. Finally, there are two comprehensive review papers on fitting a given proximity matrix (through least-squares) by a sum of matrices each having the (anti-)Robinson form (Hubert and Arabie (1994)) or the strongly-(anti-)Robinson variation (Hubert, Arabie, and Meulman (1998)). The latter discusses in detail, and with all the appropriate historical background, the need to strengthen the basic (anti-)Robinson condition to one that is strongly-(anti-)Robinson if any type of consistent graphical representation is to be achieved (and then, we might add, extends the whole graphical representation to the use of circular orders rather than the linear orders that underlie matrices having an anti-Robinson form).

[30] The program DPSE1U allows such a choice.

[31] Although we will not pursue the topic in any detail here, there are several nice theoretical relationships between these choices of a row and/or column gradient measure and the (approximate) construction of representations for the objects in S as intervals along a continuum based on 0/1 dichotomizations of the optimally reordered proximity matrix (see Roberts (1978), Chapters 3–4; Mirkin (1979), Chapter 1). In particular, the use of only one of the row or column gradient measures is relevant to the construction of (general) interval graph representations (e.g., see Mirkin (1979), Chapter 1), and there is also the use of both simultaneously in the construction of *proper* interval graphs, i.e., no interval is properly contained within another (e.g., see Roberts (1978), Chapters 3–4).

[32] Both weighted and unweighted gradient measures having this latter structure are options in the program DPSE1U and will be illustrated below.

[33] There is no approximate anti-Robinson form for the latter, however, because voters apparently do not have the same location along the continuum. If

they did, and all voters evaluated the parties in the same manner, one might expect such an anti-Robinson pattern in addition to all nonnegative above-diagonal entries.

[34]Although not an option in DPSE2U, another obvious variation would be to maximize the gradient nonviolations alone.

[35]The program DPSEPH, where the suffix 'PH' refers to 'path', that implements all these optimization variations just described has an effective limit of object set sizes of about 20, given typical current (as of 1999) RAM configurations, although as usual no formal upper-limit is built into the program.

[36]The program DPSEPH accommodates such a nonsymmetric proximity matrix in the construction of optimal linear or circular paths.

[37]The inclusion of precedence constraints through an indicator matrix is an option in each program mentioned thus far — DPSE1U, DPSE2U, and DPSEPH.

[38]The latter DHG measure is an option in the program DPSE2R, used in the numerical illustrations to follow (the suffix '2R' again denotes '2-mode restricted'). DPSE2R parallels DPSE2U in all the various options of the latter (including the possible restriction of comparisons to the rows or to the columns of either the original proximity matrix $\mathbf{Q}$ or the derived $n \times n$ matrix $\mathbf{P}^{AB}$) but allows the imposition of either column, or row and column, order constraints that can be provided by the user. It includes (as does DPSE2U) the maximization of the weighted or unweighted gradient measures of matrix patterning (and, as in DPSE2U, only the minimization of the weighted or unweighted discrepancies). As noted, when both row and column order constraints are imposed, the DHG measure just described is an additional option as well. Although again no formal limits are present in DPSE2R, practical RAM configurations will allow row object sizes of about 20 if only the (reasonably-sized) column object set is subject to an order constraint; much larger row and column object sizes (e.g., n_A's and n_B's in the hundreds) are possible when both the row and column objects are subject to order constraints.

[39]Although we will phrase our discussion as one of constructing optimal ordered partitions (or linearly ordered partitions), these are the same entities that are referred to by other names, e.g., as linear quasi-orders (Mirkin (1979), pp. 95–96), and much more commonly as weak orders (Krantz et al. (1971), pp. 14–17).

[40]The numerical illustrations rely on a program DPOP1U (where 'OP' refers to 'ordered partition' and '1U' to '1-mode unrestricted') that implements the options just described for constructing optimal ordered partitions for either symmetric or skew-symmetric matrices. The program has a built-in limit to object set sizes of 30, but given typical RAM configurations, the effective limit may actually be about 20.

Chapter 5

Heuristic Applications of the GDPP

The various applications of the GDPP in Chapters 3 and 4 to the optimization tasks of cluster analysis and object sequencing were generally limited to object sets of a certain size because of the necessary storage requirements for carrying out the attendant recursive processes. Although some possibilities may exist for reducing the extent of the basic sets, $\Omega_1, \ldots, \Omega_K$, through some type of restriction on what form an optimal solution can take, which then might allow large object sets to be approached (e.g., through linear ordering constraints), lacking such restrictions there is an inherent upper limit on the magnitude of the optimization tasks that can be handled with guaranteed optimality. If the ideal of guaranteed optimality is, for the moment, put aside, it is generally possible to use the GDPP specializations of Chapters 3 and 4 heuristically by allowing (a) the separate analyses of subsets of a (larger) object set, and (b) the use of classes of objects as the basic entities to be considered (in contrast to allowing the use of single objects only). By the judicious (and sequential) application of both these latter two options, it may be possible to analyze large object sets for the same type of clustering and sequencing tasks discussed in the last two chapters. An absolute guarantee of final optimality usually cannot be given, but we still might do quite well in producing good solutions for the optimization tasks at hand.

The two major sections of this chapter discuss the heuristic use of the GDPP within the cluster analysis context (Section 5.1) and for object sequencing and seriation (Section 5.2). The various unconstrained programs mentioned thus far in this monograph also exist in generalized forms that allow parts of a (larger) object set to be studied and the prior specification of certain object classes to be the primary units analyzed.[41] The illustrative proximity matrix considered throughout this chapter is of size 45 × 45, and refers to the dissimilarities between the 45 pairs of foods listed in Table 5.1 given at the end of this chapter (these data were kindly provided by Professors Greg Murphy and Brian Ross,

Department of Psychology, University of Illinois, Champaign, Illinois; for a different discussion and analysis of these data, see Ross and Murphy (1999)). A group of 38 subjects was asked to sort the 45 foods into as many categories as they wished, based on perceived similarity. The entries in Table 5.1 show the proportion of subjects who did not place a particular pair of foods together in a common category (thus, these proportions are keyed as dissimilarities in which larger proportions represent the more pairwise dissimilar foods). The ultimate substantive question involves the identification of the natural categories of food that may underlie the subjects' classification judgments.

The ordering given for the 45 foods in Table 5.1 (based on the analysis to be reported in Section 5.2 on sequencing these items along a continuum) allows a direct inspection of the patterning of entries and suggests that the types of categorizations given by the subjects can be diverse. For example, these might involve the differing situations in which food is used, or possibly a more basic notion of what type of food it is. For a few illustrations, 'egg' is not dramatically similar to any of the other items but does have some connection to those that involve 'breakfast' (situation), or that are 'dairy' (type), or that are 'meat' (type); 'spaghetti' appears related either to those objects that are 'entrees' and particularly to those that are 'Italian' (situation), or apparently when relying on a different interpretation for the word, to those foods that are 'cereal-based' (type); 'ice-cream' is related to 'dairy' items (type) and the 'sweet treats' given as desserts (situation).

5.1 Cluster Analysis

When faced with the task of finding a single optimal partition of a (large) object set S based on one of the heterogeneity measures/optimization criteria discussed in Section 3.1, if one somehow had knowledge that for an optimal M-class partition the classes could be allocated to two (or more) groups, then the aggregate collections of the objects within these latter groups could be optimally partitioned separately and, thus, an optimal M-class partition for the complete object set identified directly. Or, if somehow it were known that certain elemental subsets of the objects in S had to appear within the classes of an optimal M-class partition, one could begin with these elemental subsets as the objects to be analyzed, and an optimal M-class partition could again be retrieved. The obvious difficulty is that knowledge would rarely be available about either the larger aggregate groups that might be dealt with separately or an appropriate collection of elemental subsets, in a size and number that might be handled by the recursive optimization strategy discussed in Section 3.1.

Probably the best strategy is to implement these possibilities (of having aggregate collections or elemental subsets) empirically and develop a heuristic approach to analyze large object sets that would have some iterative mechanism for modifying initially conjectured aggregate collections or elemental subsets. For example, to give a heuristic strategy relying on identifying elemental subsets, one possible approach would be to begin with a partition of S into several

5.1. CLUSTER ANALYSIS

classes (possibly obtained through another heuristic process, such as complete-link hierarchical clustering), where each class contained a number of objects that could be optimally analyzed. Based on these separate subset analyses, a (tentative) collection of elemental subsets would be identified. These could then be used to obtain a subdivision of S; and again, within each group of this subdivision, the objects could be optimally partitioned to generate a possibly better collection of elemental subsets. This process could be continued until no change occurred in the particular elemental subsets identified. As an alternative, one could start with some collection of tentative elemental subsets obtained through another (heuristic) optimization strategy and try, if possible, to improve upon these through the same type of procedure. It is this latter approach that we suggest be adopted, specifically where a simple greedy strategy of hierarchical clustering is used to obtain a tentative collection of elemental subsets; the greedy heuristic will be based on the same subset heterogeneity measure optimized in finding an optimal M-class partition.

Analogously, the task of constructing a (hopefully optimal) partition hierarchy for a (larger) object set could also be approached through the identification of a collection of elemental subsets, which would then be operated on as the basic entities for the generation of a partition hierarchy. Alternatively, we could first identify a single partition, say, $\mathcal{P}_e$ (through other means), that is forced to be present in the to-be-constructed hierarchy. Beginning with the classes of $\mathcal{P}_e$, the hierarchy could first be completed optimally from that point to $\mathcal{P}_n$; each of the object subsets defined by the classes in $\mathcal{P}_e$ could then be separately, and optimally, hierarchically partitioned. The resulting collection of $n - 2$ subsets so identified will form an optimal partition hierarchy for the objects in S (and for the chosen transition measure), subject to the condition that it includes $\mathcal{P}_e$. Again, some type of iterative correction could be carried out in which subsets identified by the hierarchical partitioning of the classes in $\mathcal{P}_e$ are now aggregated (as elemental subsets) to a possibly different partition $\mathcal{P}'_e$. This process then could be repeated until no change occurs in the identification of the partition we force to be present in the hierarchy.[42]

A Partitioning Illustration:

To illustrate the heuristic approach to clustering, we choose as a subset heterogeneity measure the diameter of a cluster (i.e., the measure labeled as (iv) in Section 3.1) and adopt the optimization criterion of minimizing the maximum diameter over the classes of an M-class partition. The process suggested above for finding a tentative set of elemental object classes was carried out using a greedy selection, according to diameter, of the successive partitions in a hierarchy up to 16 classes (i.e., here, because of the choice of the diameter as the heterogeneity measure, the well-known complete-link hierarchical clustering strategy is being carried out). The 16 classes so obtained were then used in the manner described above to attempt an identification of an even better collection (i.e., grouping the 16 classes into a smaller number of subsets and then (re)partitioning these subsets). In this instance, no changes occurred in the initial collection of 16 elemental classes identified by the original greedy heuristic. We give these elemental classes below along with their diameters and suggestive interpretative labels whenever they contain more than a single object. We note that because all diameters for these 16 subsets are not larger than .50, at least 50% of the subjects placed each specific food pair contained within an elemental subset into a common class in their own sortings of the items.

Class	Possible Class Label	Diameter	Members
A	(meat/entree)	.50	lobster, hamburger, pork, steak, salmon, chicken
B	(fruit)	.08	banana, apple, watermelon, orange, pineapple
C	(liquid)	.16	water, soda
D	(non-treat dairy)	.18	butter, cheese
E	(snack)	.24	crackers, pretzels
F	(vegetable)	.24	carrots, onions, corn, lettuce, potato, broccoli
G	(sweet treat/dessert)	.26	cookies, cake, pie, chocolate bar
H	(Italian entree)	.50	pizza, spaghetti
I	(breakfast/grain-based)	.29	pancake, cereal, oatmeal, muffin, bagel
J	(grain-based)	.42	rice, bread
K	(junk food)	.45	popcorn, nuts, potato chips
L	(treat dairy)	.50	yogurt, ice cream
M	—	—	doughnuts
N	—	—	milk
O	—	—	eggs
P	—	—	granola bar

Based on these 16 elemental subsets, we provide an optimal 8-class partition (using the alphabetic labels given above) having a maximum subset diameter of .76. Again, suggestive interpretive labels are provided along with the diameters for each of the classes.

Class	Possible Class Label	Diameter
{I,J}	grain-based	.66
{H}	Italian entrees	.50
{G,M}	sweets	.61
{E,K,P}	munchies	.76
{D,L}	dairy	.63
{C,N}	liquid	.63
{B,F}	plant-based	.76
{A,O}	animal-based	.74

5.1. CLUSTER ANALYSIS

(We might note that the complete-link greedy heuristic produced a partition at the level of 8 classes [with the same elemental classes] that had a maximum diameter of .90; so, obviously, this latter partition would be nonoptimal at this level since the maximum diameter over the classes of a partition was used as the loss criterion.)

A Hierarchical Clustering Illustration:

To give an example of the heuristic use of hierarchical clustering on the data matrix of Table 5.1, we will (again) choose the diameter of a new subset formed at a given level as the measure of transition cost between partitions, and attempt to find a full partition hierarchy that would (hopefully) minimize the sum of the diameters over the $45-2 = 43$ new subsets formed in the process. To place this task in a computationally feasible framework according to the size of the object sets or object set classes that can be handled, it will be assumed that the specific 8-class partition just given in the partitioning illustration must be part of the hierarchy to be constructed. Thus, we proceed optimally from this specific 8-class partition to the trivial partition where all objects are united into a common class, and provide optimal partition hierarchies restricted to those objects within each of the 8 classes. Subject to the presence of the specific 8-class partition as part of the full partition hierarchy, this procedure guarantees an optimal hierarchy minimizing the sum of the diameters over all the subsets formed. We give the results of this process below, beginning with the 8-class partition and continuing to the single all-inclusive object class (for convenience, the same alphabetic labels are used for the 16 elemental subsets given in the partitioning application; also, the diameters of the newly formed subsets are provided along with interpretive labels whenever such appear substantively possible).

Level	Partition	Diameter [and Label]
8	(all together)	1.0
7	{IJ,EKP,H,AO,GM,DL,CN},{BF}	1.0
6	{IJ,EKP,H,AO,GM,DL},{CN},{BF}	1.0
5	{IJ,EKP,H,AO},{GM,DL},{CN},{BF}	1.0
4	{IJ,EKP},{H,AO},{GM,DL},{CN},{BF}	.95 sweets/dairy
3	{IJ,EKP},{H,AO},{GM},{DL},{CN},{BF}	.95 dinner entrees
2	{IJ,EKP},{H},{GM},{DL},{CN},{BF},{AO}	.92 grain-based
1	(all separate)	—

The optimal (sub)hierarchies for each class of the initial 8-class partition are given below, along with the diameters of the new subsets formed (generally, single object subsets are not given at any level). There are some subtle details of interpretation that might be given for several of the (sub)hierarchies but these are left to the reader's perusal (e.g., see the structure of the meat-based group).

CHAPTER 5. HEURISTIC APPLICATIONS OF GDPP

Class	Level	Partition	Diameter
{I,J}	7	(all together)	.66
	6	{bread, rice}	.42
	5	{cereal, oatmeal, pancake, muffin, bagel}	.29
	4	{cereal, oatmeal, pancake, muffin}	.18
	3	{pancake, muffin}	.16
	2	{cereal, oatmeal}	.05
	1	(all separate)	—
{H}	2	{pizza, spaghetti}	.50
	1	{pizza},{spaghetti}	—
{G,M}	5	{cake, pie, cookies, chocolate bar, doughnuts}	.61
	4	{cake, pie, cookies, chocolate bar}	.26
	3	{cookies, chocolate bar}	.16
	2	{cake, pie}	.08
	1	(all separate)	—
{E,K,P}	6	(all together)	.66
	5	{crackers, granola bar}	.47
	4	{nuts, popcorn, pretzels, potato chips}	.45
	3	{popcorn, pretzels, potato chips}	.26
	2	{pretzels, potato chips}	.24
	1	(all separate)	—
{D,L}	4	(all together)	.63
	3	{yogurt, ice cream}	.50
	2	{butter, cheese}	.18
	1	(all separate)	—
{C,N}	3	{water, soda, milk}	.63
	2	{water, soda}	.16
	1	(all separate)	—
{B,F}	11	(all together)	.76
	10	{carrots, broccoli, corn, lettuce, onions, potato}	.24
	9	{carrots, broccoli, corn, lettuce, onions}	.13
	8	{banana, apple, watermelon, orange, pineapple}	.08
	7	{carrots, broccoli, corn, lettuce}	.08
	6	{carrot, broccoli, corn}	.05
	5	{banana, apple, watermelon, orange}	.05
	4	{apple, watermelon, orange}	.03
	3	{carrots, broccoli}	.03
	2	{apple, watermelon}	.00
	1	(all separate)	—
{A,O}	7	{lobster, salmon, pork, chicken, hamburger, steak, eggs}	.74
	6	{lobster, salmon, pork, chicken, hamburger, steak}	.50
	5	{hamburger, steak, pork, chicken}	.29
	4	{hamburger, steak}	.13
	3	{pork, chicken}	.05
	2	{lobster, salmon}	.03
	1	(all separate)	—

In total, the sum of the diameters over the 43 subsets that make up the full hierarchy for the 45 food items is 17.55, which is the optimal value for any hierarchy that includes the given 8-class partition.

5.2 Object Sequencing and Seriation

The task of finding an optimal order for a (large) object set S along a continuum, based on a measure of matrix patterning discussed in Section 4.1, can be approached similarly to the large clustering problem in Section 5.1, by generalizing the GDPP recursive process to deal both with separate subsets of S and with classes of objects as the basic entities to be sequenced. If one knew, for example, that in an optimal order the object set S could be subdivided into two (or more) groups of consecutive objects, then the overall optimal order could be retrieved merely by separately and optimally sequencing the objects within each of these larger groups and treating the remaining groups as aggregate object classes to be sequenced along with the individual objects. Or, alternatively, if one had knowledge of a (larger) collection of elemental classes where within each such class the objects would be consecutively placed, the elemental classes could first be sequenced optimally and then within each of the elemental classes.[43]

A Sequencing Illustration:

To provide an example of the heuristic approach to seriation, we will choose the measure of matrix pattern based on coordinate representation and attempt to minimize the index in (4.6) for the 45×45 matrix of Table 5.1. Explicitly, the process followed involved first sequencing the 16 elemental subsets identified in Section 5.1 as classes along a continuum (with the order of the objects within each class given arbitrarily). Based on this initial order of 45 objects, several object sets (of size 18) consecutive within this current order were considered as separate objects, and those objects both before and after this consecutive set as two classes to be sequenced optimally along with all the single objects in the consecutive set (we note that just a single before or after set is considered whenever the consecutively-placed objects form an initial or an ending string). Only three such operations were necessary before an apparently optimal order was identified, which was used to provide the specific object listing for the food items in the original data in Table 5.1. (Explicitly, based on the initial order for the 16 elemental classes, the first 18 objects plus the last 27 were considered as a single class; the last 18 objects and the first 27 were considered as a single class; and a middle 18 with those objects before and after were considered as two separate classes.) Although no absolute guarantee of optimality can be given for this order, we note that it is one that cannot be improved upon by sequencing optimally any consecutive sequence of 18 objects when considering those before and after as separate classes.

The coordinates estimated for the 45 objects in the process of maximizing the index in (4.6) are given below for each of the food items in Table 5.1 (the order given for the foods in Table 5.1 is consistent with this set of coordinates):

apple	−.83	cereal	−.28	pizza	+.32
watermelon	−.83	muffin	−.25	ice cream	+.37
orange	−.82	pancake	−.24	yogurt	+.41
banana	−.82	spaghetti	−.17	butter	+.45
pineapple	−.81	crackers	−.13	cheese	+.48
lettuce	−.66	granola bar	−.10	eggs	+.53
broccoli	−.65	pretzels	−.04	milk	+.57
carrots	−.65	popcorn	+.00	water	+.63
corn	−.65	nuts	+.02	soda	+.64
onions	−.63	potato chips	+.06	hamburger	+.78
potato	−.59	doughnuts	+.12	steak	+.80
rice	−.42	cookies	+.21	pork	+.81
bread	−.34	cake	+.23	chicken	+.82
bagel	−.30	chocolate bar	+.25	lobster	+.86
oatmeal	−.29	pie	+.27	salmon	+.86

In contrast with the results from the clustering illustrations, there is a much greater amount of fine detail and subtlety available from direct inspection of the entries in Table 5.1 using this specific order as a guide, as well as several observations that can be made about which collections of entries appear to violate an approximate anti-Robinson form for the reordered proximity matrix, and why this might be so. We give some of these comments below about the patterning of entries in Table 5.1 and invite the reader to carry out a similar inspection in parallel to our conjectures:

The ordering of the food items in Table 5.1 displays a clear progression from those that are plant-based (at the top) to those that are animal-based (at the bottom). In moving from the top to the bottom, we first progress from a 'fruit' to a 'vegetable' group, where these groups both have a high degree of internal similarity and some elevated similarity within their aggregate. The item 'rice', and to some extent 'potato', placed at the end of the 'vegetable' class, can be considered spanning objects with elevated similarity back to 'vegetables' and forward to the long extent of foods that can be considered 'cereal or grain-based'. Next, there is a 'bread-stuff' pair of 'bread' and 'bagel', with a succeeding 'breakfast' subgroup (i.e., bagel, oatmeal, cereal, muffin, and pancake) in the larger expanse of 'grain-based' items. There are some interpretable anomalies in the proximities of the 'breakfast' group to two other items placed elsewhere, i.e., 'doughnuts' appears within the 'sweet treat' class and 'egg' is inside a 'dairy' grouping, but both are also possible 'breakfast' items. Continuing on with the 'grain-based' part of the ordering, we reach 'snacks/junk food' with the precursor of 'spaghetti' sitting among the 'grain-based' items

5.2. OBJECT SEQUENCING AND SERIATION

but with some understandable connections to the later-placed 'pizza' and the last six 'meat/entrees'. Note the 'granola bar' item and its small connections back to the 'breakfast' grouping including 'egg', although it appears among the 'snacks/junk food' grouping; also, 'nuts' has some small elevated and plausible similarity back to the beginning 'fruit' and 'vegetable' classes. Moving on, we reach, in order, those items that can be labeled 'sweet treats', 'dairy', 'liquid', and 'meat/entrees'. Although 'pizza' is placed within the 'treats', there is some obvious connection forward to the 'meat/entrees' category. 'Ice-cream' may be another natural spanning item, with connections back to 'sweet treats' and forward to 'dairy'; similarly, 'eggs' is placed within 'dairy' but there is some elevated similarity to 'meat/entrees' and (as mentioned earlier) back to the 'breakfast' grouping.

Endnotes

[41] For convenience, these program extensions will be referred to by replacing the 'D' (for example, in DPCL1U) by an 'H' to indicate 'heuristic', e.g., HPCL1U.

[42] Within the cluster analysis framework there are four programs that allow subsets of an object set to be analyzed or classes to be the basic objects analyzed. Three are direct extensions of DPCL1U, DPCL2U, and DPHI1U, and are called, respectively, HPCL1U, HPCL2U, and HPHI1U. A fourth, HPHI2U, is an extension of the hierarchical clustering options of HPHI1U to two-mode proximity data (and as noted in Section 3.2, would be provided at this point). In the illustrations, HPCL1U and HPHI1U have been used on the matrix of Table 5.1 for one representative heterogeneity (transition) measure available as an option in each program.

[43] Three of the programs discussed for the sequencing context (i.e., DPSE1U, DPSE2U, and DPOP1U) have been extended to allow object classes to be optimally sequenced (to HPSE1U, HPSE2U, and HPOP1U); HPSE1U is used in the numerical illustration.

Table 5.1: *A symmetric proximity matrix constructed for 45 foods based on the proportions of a group of (38) subjects who did not place a pair of foods together within a common category. (Data provided by Professors Greg Murphy and Brian Ross.)*

Foods	1	2	3	4	5	6	7	8	9	10	11	12	3	14	15
apple (1)	x	.00	.03	.05	.08	.66	.68	.71	.68	.74	.76	.87	.95	.92	.95
watermelon (2)	.00	x	.03	.05	.08	.66	.68	.71	.68	.74	.76	.87	.95	.92	.95
orange (3)	.03	.03	x	.05	.05	.68	.71	.68	.71	.74	.74	.89	.92	.95	.95
banana (4)	.05	.05	.05	x	.05	.66	.68	.68	.68	.68	.76	.89	.95	.95	.92
pineapple (5)	.08	.08	.05	.05	x	.66	.68	.66	.68	.71	.74	.89	.92	.95	.95
lettuce (6)	.66	.66	.68	.66	.66	x	.05	.08	.08	.10	.24	.74	.92	.92	.95
broccoli (7)	.68	.68	.71	.68	.68	.05	x	.03	.03	.10	.21	.71	.92	.92	.95
carrots (8)	.71	.71	.68	.68	.66	.08	.03	x	.05	.10	.18	.74	.89	.95	.95
corn (9)	.68	.68	.71	.68	.68	.08	.03	.05	x	.13	.18	.68	.92	.92	.95
onions (10)	.74	.74	.74	.68	.71	.10	.10	.10	.13	x	.21	.74	.87	.92	.89
potato (11)	.76	.76	.74	.76	.74	.24	.21	.18	.18	.21	x	.58	.79	.84	.84
rice (12)	.87	.87	.89	.89	.89	.74	.71	.74	.68	.74	.58	x	.42	.53	.53
bread (13)	.95	.95	.92	.95	.92	.92	.92	.89	.92	.87	.79	.42	x	.29	.45
bagel (14)	.92	.92	.95	.95	.95	.92	.92	.95	.92	.92	.84	.53	.29	x	.18
oatmeal (15)	.95	.95	.95	.92	.95	.95	.95	.95	.95	.89	.84	.53	.45	.18	x
cereal (16)	.95	.95	.92	.95	.92	.95	.95	.92	.95	.92	.82	.55	.40	.16	.05
muffin (17)	.97	.97	.97	.95	.97	.97	.97	.97	.97	.92	.87	.66	.45	.24	.29
pancake (18)	.97	.97	.97	.95	.97	.97	.97	.97	.97	.95	.89	.66	.50	.26	.18
spaghetti (19)	.92	.92	.95	.95	.95	.92	.92	.95	.92	.95	.87	.55	.55	.55	.63
crackers (20)	.95	.95	.92	.95	.92	.95	.95	.92	.95	.92	.82	.61	.53	.58	.61
granola bar (21)	.97	.97	.97	.95	.97	.97	.97	.97	.97	.95	.92	.74	.61	.50	.47
pretzels (22)	.97	.97	.95	.97	.95	.97	.97	.95	.97	.97	.92	.74	.71	.76	.79
popcorn (23)	.97	.97	1.0	1.0	1.0	.97	.97	1.0	.97	1.0	.97	.79	.84	.79	.79
nuts (24)	.89	.89	.89	.84	.87	.89	.89	.89	.89	.87	.92	.87	.87	.92	.87
potato chips (25)	.97	.97	.95	.97	.95	.95	.95	.92	.95	.95	.92	.87	.89	.92	.92
doughnuts (26)	1.0	1.0	1.0	.97	1.0	1.0	1.0	1.0	1.0	.97	1.0	.92	.76	.58	.55
cookies (27)	1.0	1.0	.97	1.0	.97	1.0	1.0	.97	1.0	1.0	.97	.97	.92	.95	.97
cake (28)	1.0	1.0	1.0	.97	1.0	1.0	1.0	1.0	1.0	.97	1.0	.97	.92	.92	.92
chocolate bar (29)	1.0	1.0	.97	1.0	.97	1.0	1.0	.97	1.0	1.0	.97	1.0	.97	1.0	1.0
pie (30)	1.0	1.0	1.0	.97	1.0	1.0	1.0	1.0	1.0	.97	1.0	1.0	1.0	1.0	.97
pizza (31)	.95	.95	.97	.97	.97	.95	.95	.97	.95	.97	.95	.89	.95	.95	.97
ice cream (32)	.97	.97	1.0	1.0	1.0	.97	.97	1.0	.97	1.0	1.0	.97	1.0	.97	1.0
yogurt (33)	.95	.95	.95	.92	.95	.95	.95	.95	.95	.92	.95	.95	.95	.92	.89
butter (34)	.95	.95	.97	.97	.97	.92	.92	.95	.92	.92	.95	.87	.89	.95	.97
cheese (35)	.95	.95	.97	.97	.97	.92	.92	.95	.92	.92	.95	.89	.92	.95	.97
eggs (36)	.97	.97	.97	.95	.97	.97	.97	.97	.97	.95	.97	.97	.92	.76	.66
milk (37)	1.0	1.0	.97	1.0	.97	1.0	1.0	.97	1.0	1.0	.97	1.0	.92	.95	.95
water (38)	.92	.92	.89	.92	.89	.89	.89	.87	.89	.92	.89	.95	.92	.97	.97
soda (39)	1.0	1.0	.97	1.0	.97	.97	.97	.95	.97	.97	.95	1.0	.97	1.0	1.0
hamburger (40)	.97	.97	1.0	1.0	1.0	.97	.97	1.0	.97	1.0	.97	.92	.97	.97	1.0
steak (41)	1.0	1.0	.97	1.0	.97	1.0	1.0	.97	1.0	1.0	.92	.95	.95	1.0	1.0
pork (42)	.95	.95	.97	.97	.97	.95	.95	.97	.95	.97	.92	.89	.95	.95	.97
chicken (43)	.95	.95	.97	.97	.97	.95	.95	.97	.95	.97	.92	.89	.95	.95	.97
lobster (44)	.97	.97	.97	.95	.97	.97	.97	.97	.97	.95	.95	.95	.97	.97	.95
salmon (45)	.97	.97	.97	.95	.97	.97	.97	.97	.97	.95	.95	.95	.97	.97	.95

5.2. OBJECT SEQUENCING AND SERIATION

Table 5.1: Continued.

Foods	16	17	18	19	20	21	22	23	24	25	26	27	28	29	30
apple (1)	.95	.97	.97	.92	.95	.97	.97	.97	.89	.97	1.0	1.0	1.0	1.0	1.0
watermelon (2)	.95	.97	.97	.92	.95	.97	.97	.97	.89	.97	1.0	1.0	1.0	1.0	1.0
orange (3)	.92	.97	.97	.95	.92	.97	.95	1.0	.89	.95	1.0	.97	1.0	.97	1.0
banana (4)	.95	.95	.95	.95	.95	.95	.97	1.0	.84	.97	.97	1.0	.97	1.0	.97
pineapple (5)	.92	.97	.97	.95	.92	.97	.95	1.0	.87	.95	1.0	.97	1.0	.97	1.0
lettuce (6)	.95	.97	.97	.92	.95	.97	.97	.97	.89	.95	1.0	1.0	1.0	1.0	1.0
broccoli (7)	.95	.97	.97	.92	.95	.97	.97	.97	.89	.95	1.0	1.0	1.0	1.0	1.0
carrots (8)	.92	.97	.97	.95	.92	.97	.95	1.0	.89	.92	1.0	.97	1.0	.97	1.0
corn (9)	.95	.97	.97	.92	.95	.97	.97	.97	.89	.95	1.0	1.0	1.0	1.0	1.0
onions (10)	.92	.92	.95	.95	.92	.95	.97	1.0	.87	.95	.97	1.0	.97	1.0	.97
potato (11)	.82	.87	.89	.87	.82	.92	.92	.97	.92	.92	1.0	.97	1.0	.97	1.0
rice (12)	.55	.66	.66	.55	.61	.74	.74	.79	.87	.87	.92	.97	.97	1.0	1.0
bread (13)	.40	.45	.50	.55	.53	.61	.71	.84	.87	.89	.76	.92	.92	.97	1.0
bagel (14)	.16	.24	.26	.55	.58	.50	.76	.79	.92	.92	.58	.95	.92	1.0	1.0
oatmeal (15)	.05	.29	.18	.63	.61	.47	.79	.79	.87	.92	.55	.97	.92	1.0	.97
cereal (16)	x	.29	.18	.61	.55	.47	.74	.82	.92	.89	.55	.92	.92	.97	1.0
muffin (17)	.29	x	.16	.68	.58	.53	.74	.79	.89	.87	.45	.92	.87	1.0	.95
pancake (18)	.18	.16	x	.66	.66	.58	.79	.84	.92	.89	.42	.89	.84	.97	.92
spaghetti (19)	.61	.68	.66	x	.66	.71	.76	.79	.92	.95	.92	.95	.92	1.0	1.0
crackers (20)	.55	.58	.66	.66	x	.47	.24	.40	.58	.45	.89	.82	.92	.87	1.0
granola bar (21)	.47	.53	.58	.71	.47	x	.53	.50	.63	.66	.71	.74	.76	.76	.84
pretzels (22)	.74	.74	.79	.76	.24	.53	x	.26	.45	.24	.82	.76	.89	.82	.92
popcorn (23)	.82	.79	.84	.79	.40	.50	.26	x	.42	.26	.87	.74	.84	.76	.89
nuts (24)	.92	.89	.92	.92	.58	.63	.45	.42	x	.45	.89	.84	.87	.84	.89
potato chips (25)	.89	.87	.89	.95	.45	.66	.24	.26	.45	x	.79	.66	.79	.63	.76
doughnuts (26)	.55	.45	.42	.92	.89	.71	.82	.87	.89	.79	x	.53	.53	.61	.55
cookies (27)	.92	.92	.89	.95	.82	.74	.76	.74	.84	.66	.53	x	.18	.16	.21
cake (28)	.92	.87	.84	.92	.92	.76	.89	.84	.87	.79	.53	.18	x	.26	.08
chocolate bar (29)	.97	1.0	.97	1.0	.87	.76	.82	.76	.84	.63	.61	.16	.26	x	.21
pie (30)	1.0	.95	.92	1.0	1.0	.84	.92	.89	.89	.76	.55	.21	.08	.21	x
pizza (31)	.97	1.0	.97	.50	.89	.82	.79	.74	.79	.68	.76	.74	.76	.71	.68
ice cream (32)	1.0	1.0	.97	.97	.97	.92	.92	.87	.89	.82	.76	.50	.40	.47	.40
yogurt (33)	.92	.87	.92	.92	.84	.82	.84	.84	.82	.89	.92	.89	.82	.87	.82
butter (34)	.97	1.0	.97	.95	.97	.92	.89	.87	.89	.82	.84	.82	.82	.82	.82
cheese (35)	.97	1.0	.97	.95	.95	.92	.92	.92	.87	.92	.95	.95	.95	.95	.95
eggs (36)	.68	.79	.68	.95	.97	.79	.92	.95	.79	.97	.74	.97	.95	.97	.95
milk (37)	.92	.95	.95	.97	.97	.92	.92	.95	.97	.95	.92	.95	.97	.95	.97
water (38)	.95	.97	.97	.97	.92	.95	.95	.97	1.0	.95	.97	.92	.95	.89	.95
soda (39)	.97	1.0	.97	1.0	.97	.92	.89	.92	.95	.82	.87	.79	.84	.76	.82
hamburger (40)	1.0	1.0	.97	.71	.97	.97	.95	.92	.87	.92	.95	.95	.95	.92	.92
steak (41)	.97	1.0	.97	.76	.97	.97	.95	.97	.89	.92	.95	.92	.95	.92	.95
pork (42)	.97	.97	.95	.71	.97	1.0	1.0	.97	.89	.97	.97	.97	.97	.97	.97
chicken (43)	.97	.97	.97	.68	.97	.97	.97	.95	.92	1.0	1.0	1.0	1.0	1.0	1.0
lobster (44)	.97	.95	.92	.76	.97	.97	1.0	1.0	.89	.97	.95	.97	.95	.97	.95
salmon (45)	.97	.95	.95	.74	.97	.95	.97	.97	.92	1.0	.97	1.0	.97	1.0	.97

Table 5.1: Continued.

Foods	31	32	33	34	35	36	37	38	39	40	41	42	43	44	45
apple (1)	.95	.97	.95	.95	.95	.97	1.0	.92	1.0	.97	1.0	.95	.95	.97	.97
watermelon (2)	.95	.97	.95	.95	.95	.97	1.0	.92	1.0	.97	1.0	.95	.95	.97	.97
orange (3)	.97	1.0	.95	.97	.97	.97	.97	.89	.97	1.0	.97	.97	.97	.97	.97
banana (4)	.97	1.0	.92	.97	.97	.95	1.0	.92	1.0	1.0	1.0	.97	.97	.95	.95
pineapple (5)	.97	1.0	.95	.97	.97	.97	.97	.89	.97	1.0	.97	.97	.97	.97	.97
lettuce (6)	.95	.97	.95	.92	.92	.97	1.0	.89	.97	.97	1.0	.95	.95	.97	.97
broccoli (7)	.95	.97	.95	.92	.92	.97	1.0	.89	.97	.97	1.0	.95	.95	.97	.97
carrots (8)	.97	1.0	.95	.95	.95	.97	.97	.87	.95	1.0	.97	.97	.97	.97	.97
corn (9)	.95	.97	.95	.92	.92	.97	1.0	.89	.97	.97	1.0	.95	.95	.97	.97
onions (10)	.97	1.0	.92	.92	.92	.95	1.0	.92	.97	1.0	1.0	.97	.97	.95	.95
potato (11)	.95	1.0	.95	.95	.95	.97	.97	.89	.95	.97	.92	.92	.92	.95	.95
rice (12)	.89	.97	.95	.87	.89	.97	1.0	.95	1.0	.92	.95	.89	.89	.95	.95
bread (13)	.95	1.0	.95	.89	.92	.92	.92	.92	.97	.97	.95	.95	.95	.97	.97
bagel (14)	.95	.97	.92	.95	.95	.76	.95	.97	1.0	.97	1.0	.95	.95	.97	.97
oatmeal (15)	.97	1.0	.89	.97	.97	.66	.95	.97	1.0	1.0	1.0	.97	.97	.95	.95
cereal (16)	.97	1.0	.92	.97	.97	.68	.92	.95	.97	1.0	.97	.97	.97	.97	.97
muffin (17)	1.0	1.0	.87	1.0	1.0	.79	.95	.97	1.0	1.0	1.0	.97	.97	.95	.95
pancake (18)	.97	.97	.92	.97	.97	.68	.95	.97	.97	.97	.97	.95	.97	.92	.95
spaghetti (19)	.50	.97	.92	.95	.95	.95	.97	.97	1.0	.71	.76	.71	.68	.76	.74
crackers (20)	.89	.97	.84	.97	.95	.97	.97	.92	.97	.97	.97	.97	.97	.97	.97
granola bar (21)	.82	.92	.82	.92	.92	.79	.92	.95	.92	.97	.97	1.0	.97	.97	.95
pretzels (22)	.79	.92	.84	.89	.92	.92	.92	.95	.89	.95	.95	1.0	.97	1.0	.97
popcorn (23)	.74	.87	.84	.87	.92	.95	.95	.97	.92	.92	.97	.97	.95	1.0	.97
nuts (24)	.79	.89	.82	.89	.87	.79	.97	1.0	.95	.87	.89	.89	.92	.89	.92
potato chips (25)	.68	.82	.89	.82	.92	.97	.95	.95	.82	.92	.92	.97	1.0	.97	1.0
doughnuts (26)	.76	.76	.92	.84	.95	.74	.92	.97	.87	.95	.95	.97	1.0	.95	.97
cookies (27)	.74	.50	.89	.82	.95	.97	.95	.92	.79	.95	.92	.97	1.0	.97	1.0
cake (28)	.76	.40	.82	.82	.95	.95	.97	.95	.84	.95	.95	.97	1.0	.95	.97
chocolate bar (29)	.71	.47	.87	.82	.95	.97	.95	.89	.76	.92	.92	.97	1.0	.97	1.0
pie (30)	.68	.40	.82	.82	.95	.95	.97	.95	.82	.92	.95	.97	1.0	.95	.97
pizza (31)	x	.87	.92	.79	.89	.95	.97	.95	.87	.61	.71	.74	.76	.79	.82
ice cream (32)	.87	x	.50	.63	.61	.87	.68	1.0	.89	.92	.95	.95	.97	.97	1.0
yogurt (33)	.92	.50	x	.53	.42	.68	.53	.92	1.0	1.0	1.0	.97	.95	.95	.92
butter (34)	.79	.63	.53	x	.18	.74	.58	.97	.89	.92	.95	.95	.97	.97	1.0
cheese (35)	.89	.61	.42	.18	x	.66	.45	.97	.95	.92	.95	.95	.97	.97	1.0
eggs (36)	.95	.87	.68	.74	.66	x	.71	1.0	.97	.74	.71	.71	.71	.74	.71
milk (37)	.97	.68	.53	.58	.45	.71	x	.61	.63	.97	.95	1.0	.97	1.0	.97
water (38)	.95	1.0	.92	.97	.97	1.0	.61	x	.16	1.0	.97	.97	.97	.97	.97
soda (39)	.87	.89	1.0	.89	.95	.97	.63	.16	x	.95	.92	.97	1.0	.97	1.0
hamburger (40)	.61	.92	1.0	.92	.92	.74	.97	1.0	.95	x	.13	.24	.29	.47	.50
steak (41)	.71	.95	1.0	.95	.95	.71	.95	.97	.92	.13	x	.16	.21	.45	.47
pork (42)	.74	.95	.97	.95	.95	.71	1.0	.97	.97	.24	.16	x	.05	.34	.37
chicken (43)	.76	.97	.95	.97	.97	.71	.97	.97	1.0	.29	.21	.05	x	.34	.32
lobster (44)	.79	.97	.95	.97	.97	.74	1.0	.97	.97	.47	.45	.34	.34	x	.03
salmon (45)	.82	1.0	.92	1.0	1.0	.71	.97	.97	1.0	.50	.47	.37	.32	.03	x

Chapter 6

Extensions and Generalizations

6.1 Introduction

There are a variety of extensions of the topics introduced in the previous chapters that could be pursued, several of which have been mentioned earlier along with a comment that they would not be developed in any detail within this monograph. Among some of these possibilities are: (a) the development of a mechanism for generating all the optimal solutions for a specific optimization task when multiple optima may be present, not just one representative exemplar; (b) the incorporation of other loss or merit measures within the various sequencing and partitioning contexts discussed; (c) extensions to the analysis of arbitrary t-mode data, with possible (order) restrictions on some modes but not others, or to a framework in which proximity is given on more than just a pair of objects, e.g., proximity could be defined for all distinct object triples (see Daws (1996)); (d) the generalization of the task of constructing optimal ordered partitions to a two- or higher-mode context that may be hierarchical and/or have various types of order or precedence constraints imposed; and (e) the extension of object ordering constraints when they are to be imposed (e.g., in various partitioning and two-mode sequencing tasks) to the use of circular object orders, where optimal subsets or ordered sequences must now be consistent with respect to a circular contiguity structure.

In this last chapter, we note four other general areas in which more work could be pursued: (a) the incorporation of multiple sources of data about the objects under study; (b) the identification of multiple structures for the representation of a particular data set; (c) a suggestion that the information produced while carrying out a recursive process and stored for the entities in the sets $\Omega_1, \ldots, \Omega_K$ may be of value both diagnostically and for obtaining a better substantive understanding of the objects under study by identifying alternative combinatorial structures that although not necessarily optimal, may nevertheless be very close; and (d) the use of a priori nonnegative weights on the objects, typically indicating identical replication, or dichotomous (0/1) weights on the proximities specifying missing information. We will discuss briefly, in turn, each of these four topics.

6.1.1 Multiple Data Sources

The various specializations of the GDPP introduced in the previous chapters all had the characteristic that the data guiding the optimization process are given as a *single* proximity matrix, whether symmetric, skew-symmetric, or two-mode. If, instead, a collection of, say, N such proximity matrices are available (i.e., a three-way data set, in the terminology of Tucker (1964)), and we seek to use all N sources in finding an optimal structure, the attendant recursive processes could easily be modified to do so. For example, in the clustering context, whether hierarchical or not, the main computational aspects of the recursions involve the calculation of the subset heterogeneity measures. For a given subset, these could first be defined separately for each of the N proximity matrices, and an aggregate of the latter (such as a simple average or a maximum, or alternatively, some type of weighted average based either on a set of a priori weights or those identified through a process of optimal scaling) could then be used to obtain the heterogeneity measure needed in carrying out the recursive process. Once an optimal result has been obtained based on the subset heterogeneity measures aggregated over the N data sources, each of the latter could then be evaluated individually against this common outcome. For example, the adequacy of an overall clustering solution in explaining the patterning of the data for each source could be assessed through the same type of measure of total cost adopted for the recursive process, i.e., minimizing the sum or the maximum heterogeneity measure over subsets. Depending on the particular heterogeneity measure chosen, it even may be possible to bypass this more involved construction of the overall subset heterogeneity measures by the construction of a single proximity matrix that itself is defined through a process of aggregation over the N proximity matrices.

For example, suppose the diameter criterion (i.e., the measure in (iv) from Section 3.1) is selected and subset heterogeneity is to be the maximum of the N diameters over the data sources. A single proximity matrix in which an entry for a specific object pair is itself the maximum of the N proximities for that pair would suffice in this case. Similarly, if the average of the average within-subset proximities over the N sources is taken as the measure of subset heterogeneity, then only a single proximity matrix would be needed with entries defined by the average proximity for each specific object pair over the N sources. Analogously, all the various sequencing tasks could incorporate multiple data sources by finding the contribution for the placement of an object (or for an object class when constructing optimal ordered partitions) by aggregating the contributions over the N data sources. Again, how well a final (common) order does with respect to each of the separate data sources could be assessed using the same measure

6.1. INTRODUCTION

optimized in finding the common order but now applied individually to each data source.

Although there may be no computational advantage in first reducing N data sources to a single proximity matrix and then analyzing only the latter, there are several nice theoretical observations we can make when this strategy may be possible, particularly in one special case (which follows the work of Mirkin by incorporating proximity thresholds in the clustering process, e.g., see Mirkin (1979), Chapter 2 and the discussion in Hubert, Arabie, and Meulman (1997b)). To be explicit, suppose the N data sources are (for now) N symmetric $n \times n$ proximity matrices (i.e., a two-mode three-way matrix) where each contains entries that lie between 0 and 1 (which could be strictly dichotomous and take on only the values 0 or 1), and denoted by $\mathbf{P}^{(1)}, \ldots, \mathbf{P}^{(N)}$. The optimization task to be solved is one of finding a partition of the objects in S represented by a (recovered) 0/1 proximity matrix, say, $\mathbf{P}^{(*)}$ (where an entry of 0 indicates an object pair within a class and 1 otherwise), such that a sum of N distances defined between $\mathbf{P}^{(*)}$ and $\mathbf{P}^{(1)}, \ldots, \mathbf{P}^{(N)}$ is as small as possible, i.e., we minimize

$$\sum_{h=1}^{N} d(\mathbf{P}^{(h)}, \mathbf{P}^{(*)}),$$

where the distance between $\mathbf{P}^{(h)}$ and $\mathbf{P}^{(*)}$ is defined by

$$d(\mathbf{P}^{(h)}, \mathbf{P}^{(*)}) = \sum_{i,j} \mid p_{ij}^{(h)} - p_{ij}^{(*)} \mid .$$

Note in particular that the number of object classes characterizing $\mathbf{P}^{(*)}$ is not fixed in advance and is part of the optimization task being posed.

To obtain a solution to this latter optimization problem, a single proximity matrix $\mathbf{P}$ is first constructed from the average proximity over the N matrices, $\mathbf{P}^{(1)}, \ldots, \mathbf{P}^{(N)}$, each deviated from a threshold value of $1/2$:

$$\mathbf{P} = \{p_{ij}\} = \left\{ \left(\frac{1}{N}\right) \sum_{i,j} p_{ij}^{(h)} - \left(\frac{1}{2}\right) \right\}. \tag{6.1}$$

This single dissimilarity matrix $\mathbf{P}$ is then used with the subset heterogeneity measure defined by the sum of the within-class proximities (i.e., the measure from Section 3.1 denoted as (i)) and with the optimization criterion of minimizing the sum of subset heterogeneity measures over the classes of a partition. If the minimum is found over all possible numbers of classes from 1 to n, an optimal $\mathbf{P}^{(*)}$ is identified. Because an entry in $\mathbf{P}$ is simply the average for that object pair over all N data sources minus the constant term $1/2$, this latter proximity matrix $\mathbf{P}$ together with the subset heterogeneity measure (and in contrast with using only the average proximity without this subtractive term) tends to guide $\mathbf{P}^{(*)}$ toward a set of classes that are both relatively disparate in size and few in number.

As a brief illustration of how this process might be carried out, the original sorting data for the 38 subjects who produced the food proximity matrix of Table 5.1 could be interpreted as arising from 38 dissimilarity matrices each defining a partition of the 45 foods, and the Table 5.1 proportions give $\{(1/N)\sum_{i,j} p_{ij}^{(h)}\}$. Thus, when 1/2 is first subtracted from each entry in Table 5.1, the process of finding a partition $\mathbf{P}^{(*)}$ for the 45 foods minimizing $\sum_{h=1}^{N} d(\mathbf{P}^{(h)}, \mathbf{P}^{(*)})$ can be carried out by minimizing the sum of within-class proximities defined by (6.1). As a small example using only the last 16 foods listed in Table 5.1 (and thus, given the size of the object set, an optimal DP approach can be applied), the optimal partition contains 7 classes (with a total cost index of -7.894) defined as follows: {{ice cream, pie}, {eggs}, {milk}, {soda, water}, {cheese, butter, yogurt}, {pizza}, {chicken, salmon, steak, pork, hamburger, lobster}}. All other optimal partitions with from 1 to 13 classes had total cost indices larger than -7.894. (We note that if the original proximities of Table 5.1 were used without the constant subtraction of 1/2, the sum of within-class proximities would monotonically increase for those partitions identified as optimal for 13 to 1 class(es).)

A second example of this general approach is the use of multiple proximity matrices that represent degrees of dominance among the objects in some set S. Specifically, suppose the N data sources are again $n \times n$ matrices $\mathbf{P}^{(1)}, \ldots, \mathbf{P}^{(N)}$, where each contains entries that lie between 0 and 1 but is (typically) not symmetric. The optimization task we now wish to solve is to find an ordered partition for the objects in S represented by a nonsymmetric matrix with 0/1 entries, say, $\mathbf{P}^{(**)}$ (where each ordered object pair is indicated by a 0 when the objects are either within a class of the partition or defined between two classes of the partition where the first object in the pair belongs to a class that precedes the class to which the second object belongs), that will minimize

$$\sum_{h=1}^{N} d(\mathbf{P}^{(h)}, \mathbf{P}^{(**)}).$$

If the single matrix $\mathbf{P}$ is again obtained as in (6.1) and used to find optimal ordered partitions that maximize the above-diagonal sums, the maximum over all possible numbers of classes from 1 to n is the optimal $\mathbf{P}^{(**)}$. Here, the incorporation of the negative 1/2 in (6.1) tends to guide $\mathbf{P}^{(**)}$ toward a set of ordered classes that are relatively close to each other in size and many in number.

An illustration of identifying an optimal ordered partition can be generated from the data given in Table 1.2 on the proportions of school children who rate one offense as more serious than another. Considering just the 'after' data for convenience, each child could be assumed to have produced a matrix $\mathbf{P}^{(h)}$, where an entry is 1 if the column offense is rated more serious than the row offense and 0 otherwise. Thus, the (lower-triangular) entries of Table 1.2 provide the aggregate matrix, and if 1/2 is subtracted from each entry, the proximity matrix $\mathbf{P}$ from (6.1) is skew-symmetric (with entries that are one-half of those provided in the lower-diagonal portion of Table 4.1, with the asterisks showing those that are negative). Based on this latter skew-symmetric matrix **P**, the ordered partition, maximizing the above-diagonal sum over all ordered partitions from 1 to 13 classes, contains the maximum of 13 classes (with an index value of 23.71) with its classes ordered exactly as in Section 4.1.2, i.e., $\{9\} \prec \{7\} \prec \{10\} \prec \{4\} \prec \{11\} \prec \{3\} \prec \{2\} \prec \{6\} \prec \{5\} \prec \{13\} \prec \{1\} \prec \{12\} \prec \{8\}$.

6.1.2 Multiple Structures

All the optimization tasks encountered in the earlier chapters were concerned with the identification of one optimal structure intended to help interpret the patterning of entries in a given proximity matrix, where this single structure was possibly an object partition or a partition hierarchy, or some type of object sequencing along a continuum. If some mechanism is adopted for obtaining a set of residuals (i.e., differences between the original proximities and the corresponding set of fitted proximities) based on the single optimal structure found for the original proximity matrix, these residuals can be interpreted as forming another proximity matrix. In turn, a second optimal structure can be identified that may help us interpret the patterning present in the collection of residuals, and thus a second set of fitted values can be now defined for these residuals. Together, the two optimal structures and the fitted values each produces (one based on the original proximity information and the second through the set of residuals generated from the first optimal structure) serve to explain the initial proximity values additively. Obviously, this process of residualization could be carried out more than once. Moreover, after some fixed number of optimal structures has been identified using successive residuals, the whole process could then be iterated, first by adding back to a last set of obtained residuals those predicted values from the first optimal structure identified to see if an even better first structure could now be obtained. Similarly, the procedure could be continued for the second optimal structure identified by first adding back the predicted values from this second optimal structure, and so on cyclically until hopefully some type of convergence is attained, i.e., until the same set of optimal structures reappears each time through the complete process.

The three main combinatorial structures emphasized in the preceding chapters (i.e., partitions, partition hierarchies, and object sequences of various kinds) all can be used to define appropriate collections of linear equality/inequality constraints that a set of fitted values must satisfy, irrespective of the specific choice of the L_p loss function that might be minimized (e.g., the sum of absolute differences, the sum of squared differences, or the maximum absolute deviation). Using these constraints and an appropriate L_p regression routine (e.g., one of those given in Späth (1991)), the fitted values could be obtained for either the original proximities or for a set of residuals generated from the use of other identified structures. To be explicit about the constraints implied by each of the three basic structures, we have for (a) an M-class partition: the fitted values within each of the M separate classes must be equal, and all fitted values between two specific classes must be equal; (b) a partition hierarchy: the fitted values must satisfy the linear equality/inequality constraints implied by the basic ultrametric matrix corresponding to the given partition hierarchy (as presented explicitly in Section 3.2.1); (c) an object sequencing: the fitted values must satisfy the equality/inequality constraints within rows and columns characterizing the gradient conditions given in Section 4.1.1 for an anti-Robinson form.

As a very brief illustration of such residualization, we consider the data in Table 1.1 for the pairwise dissimilarities of the ten digits and the residuals that would be generated from the 4-group partition identified most often as optimal (i.e., $\{\{0,1\},\{2,4,8\},\{3,6,9\},\{5,7\}\}$). The least-squares alternative is adopted that would define the fitted values as arithmetic averages of the within- or the between-subset proximities; these are provided below:

class	{0,1}	{2,4,8}	{3,6,9}	{5,7}
{0,1}	.421	.635	.679	.733
{2,4,8}	x	.224	.464	.615
{3,6,9}	x	x	.286	.481
{5,7}	x	x	x	.400

If the residuals from the original proximity values are generated from these fitted values and then used to obtain optimal partitions into 2 through 9 classes using the sum of the within-class sums as the cost criterion, in each case the classes of an optimal partition are defined by digits consecutive in magnitude. Thus, the structural properties of the digits appear best-represented by the first identified optimal partition constructed for the original proximities; however, at least part of any unexplained patterning in the residuals appears interpretable through numerical magnitude. We give below the optimal partitions (without listing single object sets) having 2 though 9 classes (based on the residuals) along with the cumulative sums of the within-class sums (which, as one might expect, are all negative):

Number of Classes	Partition	Cumulative Sum
2	{0,1,2,3,4,5},{6,7,8,9}	−3.448
3	{1,2,3,4,5},{6,7,8,9}	−3.646
4	{1,2,3,4,5},{6,7,8}	−3.478
5	{1,2,3,4,5},{7,8}	−3.122
6	{1,2,3,4,5}	−2.692
7	{1,2,3},{7,8}	−2.018
8	{1,2,3}	−1.588
9	{1,2}	−.702

6.1.3 Uses for the Information in the Sets $\Omega_1, \ldots, \Omega_K$

For all specializations of the GDPP presented in the previous chapters, the recursive processes were defined over a collection of sets, $\Omega_1, \ldots, \Omega_K$, and an optimal solution to the specific optimization task under consideration was identified

6.1. INTRODUCTION

from just one member contained in Ω_K. In carrying out a recursive strategy, however, a substantial amount of additional information is generated for the entities $A_k \in \Omega_k$ for $1 \leq k \leq K$, and in particular, each quantity $\mathcal{F}(A_k)$ defines an optimal value for A_k that potentially varies in its substantive interpretation depending on how the basic set Ω_k is characterized. These latter optimal values may help suggest possible alternative solutions to the optimization task being solved and be of general diagnostic assistance in explaining what the original proximities tell us substantively about the objects in the set S. We discuss briefly how this additional information stored for the sets $\Omega_1, \ldots, \Omega_K$ might be used for several GDPP specializations we have developed.

In the construction of an optimal M-class partition, the sets $\Omega_1, \ldots, \Omega_M$ were defined so each contained all subsets of an original object set S, and $\mathcal{F}(A_k)$ for $A_k \in \Omega_k$ was specified to be an optimal value for the cost criterion if just those objects in A_k were partitioned optimally into k classes. This last observation was used to construct, in addition to a single optimal M-class partition identified by $\mathcal{F}(A_M)$ for $A_M = S \in \Omega_M$, the optimal partitions of S into 2 through $M-1$ classes using $\mathcal{F}(A_k)$ when $A_k = S \in \Omega_k$ for $2 \leq k \leq M-1$. By comparing the stored values of $\mathcal{F}(A_k)$ for any set $A_k \in \Omega_k$ to $\mathcal{F}(S)$ for $S \in \Omega_k$, a direct assessment can be made as to which object(s) have the most "influence" on the total cost criterion in obtaining an optimal k-class partition of S, and thus, indirectly, as to which object(s) might be considered "outliers" or which subsets of S are particularly well-partitionable according to the chosen cost criterion. Besides identifying potentially influential object(s), the values $\mathcal{F}(A_k)$ can also suggest other good, but not necessarily optimal, solutions for the partitioning of S into $k+1$ classes. Specifically, suppose we go back to the stored values $\mathcal{F}(A_k)$ and add to each the cost increments for transforming A_k directly to S (i.e., we add the heterogeneity measures for the single subsets $S - A_k$), and order from smallest to largest the resultant sums (or alternatively, the maximum of the two terms when the cost criterion is the maximum of the heterogeneity measures over the classes of a partition). This ordered set of values identifies a collection of $(k+1)$-class partitions that may be competitive with the one identified as optimal (the optimal value for $\mathcal{F}(S)$ for $S \in \Omega_{k+1}$ will appear as the smallest in this ordered list at least k times, but those next in size reflect other $(k+1)$-class partitions that might be almost as good).

A similar evaluation procedure is possible when allowing only M-class partitions that are order-constrained or for constructing optimal ordered partitions. In this latter case, for example, the values for $\mathcal{F}(A_k)$ provide information on how well subsets of S can be partitioned into k ordered classes that are themselves sequenced along a continuum. Moreover, when the number of objects in the subset of S and the number of ordered classes are constrained to be equal, the assessment becomes one of how well the individual objects in that subset of S can be sequenced along a continuum. If these latter values are compared to the criterion obtained when the complete set S is optimally partitioned into n ordered classes, a determination can again be made about which object(s) might be considered outliers or particularly difficult to sequence along a continuum in the presence of the remaining objects in S. Also, one may be able to identify a collection of reasonably good $(k+1)$-class ordered partitions (similarly to the suggestion for finding $(k+1)$-class [unordered] partitions) formed directly from $\mathcal{F}(A_k)$ for any $A_k \in \Omega_k$ and the additional contribution needed in transforming A_k directly to S.

In the specialization of the GDPP to a hierarchical clustering task, the set Ω_k contained (for the agglomerative interpretation) all partitions of the n objects in S into $n-k+1$ classes, and thus, $\mathcal{F}(A_k)$ for $A_k \in \Omega_k$ is the optimal cost criterion that can be achieved for a partial partition hierarchy that would begin with the single trivial partition $A_1 \in \Omega_1$, in which all objects are placed into separate classes, and terminate with the specific partition represented by A_k. This observation was used previously in Section 3.2 to generate an optimal partial partition hierarchy up to $n-k+1$ classes by finding the minimal value for $\mathcal{F}(A_k)$ over all $A_k \in \Omega_k$. Besides providing a mechanism for identifying optimal partial partition hierarchies, the values for $\mathcal{F}(A_k)$ can also be used to assess directly how well subsets of S can be hierarchically partitioned, and in turn, which object(s) may be particularly difficult to include in a good partition hierarchy or which might be characterized as outliers. Specifically, an optimal partition hierarchy restricted to a subset of S containing, say, h objects, can be identified immediately from $\mathcal{F}(A_h)$, where $A_h \in \Omega_h$ is defined by that partition of S in which all h objects are in one class and the remaining $n-h$ objects are in $n-h$ separate classes. These latter optimal values for subsets of S can be compared to $\mathcal{F}(S)$ for $S \in \Omega_n$ to evaluate the effect of eliminating $n-h$ objects(s) on the construction of an optimal partition hierarchy for the complete object set S.

Much as in the M-class partition context, the values stored for some entities in the sets $\Omega_1, \ldots, \Omega_n$ may also be used more directly to identify alternative partition hierarchies that may be good but not necessarily optimal. For example, one could observe which partition hierarchies are attached to the better values obtained for the entities $A_{n-1} \in \Omega_{n-1}$ (i.e., partitions into two classes) that are not part of the optimal hierarchy, or, considering the specific entity A_{n-1} that is part of the optimal hierarchy, whether those partitions $A_{n-2} \in \Omega_{n-2}$ that are transformable into A_{n-1} are attached to partition hierarchies (other than the optimal one) that also might be good according to the total cost criterion.

For the task of sequencing objects along a continuum, which is a special case of constructing optimal ordered partitions where the number of classes in an ordered partition is now equal to the number of objects, we have already noted above how the influence of particular object(s) might be evaluated. When objects are ordered though the construction of optimal paths, we can also assess how well any subset of S can be so sequenced merely by minimizing over those entries $\mathcal{F}((A_k, j_k))$ (where it should be recalled that $(A_k, j_k) \in \Omega_k$ is now defined by an ordered pair involving a subset of S and a last-placed object from that subset) in which the particular subset is the first member of the ordered pair and one of its elements is the second. A comparison of these latter values to the total optimal value achievable using all the objects in S provides a mechanism for assessing the effect that particular object(s) may have on the total optimal value.

In sequencing objects along a continuum either through the construction of optimal paths or using a measure of matrix patterning discussed in Section 4.1, the search for good but not necessarily optimal orderings might proceed differently than in a clustering context. Here, we consider one-mode symmetric proximity matrices and implicitly assume the use of a measure of matrix patterning discussed in Section 4.1, but extensions of this general type of approach could be developed for object sequencing based on optimal path construction or for the inclusion of skew-symmetric matrices (in this latter case, some small modification would be necessary for how incremental merit contributions are defined when single objects are placed along a continuum). Explicitly, for one-mode symmetric proximity matrices, the recursive process followed for obtaining an optimal sequencing of S produces (along the way) the function values $\mathcal{F}(A_k)$ associated with $A_k \in \Omega_k$ that show a contribution to the total merit measure if the objects in A_k are placed in the first k positions (reading from left to right). Given the structure of this optimization task and the merit measure being used, these same function values $\mathcal{F}(A_k)$ would also yield the contribution to the total merit measure if the objects in A_k appeared in the last k positions. Thus, the sum of the stored values $\mathcal{F}(A_k)$ for $A_k \in \Omega_k$ and $\mathcal{F}(S - A_k)$ for $S - A_k \in \Omega_{n-k}$ defines an optimal value for an ordering of the objects in S when the objects in the set A_k appear somewhere in the first k positions and those in $S - A_k$ appear somewhere in the last $n - k$ positions. By ordering these sums from largest to smallest for all combinations of A_k and $S - A_k$, the optimal value achievable for the complete set S will appear $n - 1$ times as the largest in the list, but there may be other identified orderings that, although not necessarily optimal, might still be reasonably good.[44]

6.1.4 A Priori Weights for Objects and/or Proximities

Among the difficulties in using a specialization of the GDPP to obtain an optimal solution for some clustering or sequencing task is that we are severely limited in the absolute size of problems that can be considered. There is a context, however, in which object set sizes could be rather enormous, so long as there are only a reasonable number of *distinct* object types available and these object types define the set S to be studied. Specifically, for all the various optimization tasks discussed, and assuming identical objects must be placed together in whatever optimal combinatorial structure is being sought, all the specializations of the GDPP discussed could be generalized to incorporate positive weights (say, $w_1, \ldots, w_n$) for each of the n object types constituting the set S. For example, in the partitioning task, using the sum of within-subset proximities as the measure of subset heterogeneity, the contribution for some pair of objects, say, O_i and O_j, placed within the same subset, S_a, would be $w_i w_j p_{ij}$, where p_{ij} is the proximity between object types O_i and O_j, and the overall heterogeneity index for S_a is

$$\sum_{i,j \in S_a} w_i w_j p_{ij}.$$

If the average within-subset proximity were instead used, the heterogeneity for S_a could be given as

$$\left(\frac{1}{\sum_{i,j\in S_a} w_i w_j}\right) \sum_{i,j\in S_a} w_i w_j p_{ij}.$$

Analogously, in a sequencing context using the gradient measure in (4.1), the definitions for $\sum_{k=1}^{n} I_{row}(k)$ and $\sum_{k=1}^{n} I_{col}(k)$ in implementing the GDPP would be extended as

$$I_{row}(k) \equiv \sum_{i=1}^{k-1} \sum_{j=k+1}^{n} (w_i w_k)(w_i w_j) f(p_{ik}, p_{ij})$$

and

$$I_{col}(k) \equiv \sum_{i=1}^{k-1} \sum_{j=k+1}^{n} (w_k w_j)(w_i w_j) f(p_{kj}, p_{ij}).$$

In the heuristic extensions for partitioning, sequencing, and hierarchical clustering discussed in Chapter 5, a form of object weighting similar to that just described would result as a special case when the classes to be analyzed (as the basic objects) are defined according to whether they have identical objects. In these instances, the heuristic designation for the applications in Chapter 5 is no longer appropriate; optimality is guaranteed at the level of the a priori classes being analyzed, and these must be considered as indivisible entities. In general, however, it may be most efficient merely to incorporate object weights directly into specializations of the GDPP; the heuristic programs in Chapter 5 require the definition of the object classes on-line, and a complete proximity matrix must first be input even between replicated objects. A direct inclusion of object weights would require only the use of the proximity matrix between object types plus the set of nonnegative weights $w_1, \ldots, w_n$.

A second form of weighting that might be developed for the various specializations of the GDPP would not be on the objects in a set S but rather for the proximities defined between pairs of the objects in S. If $\mathbf{P} = \{p_{ij}\}$ denotes (as usual) the symmetric proximity matrix between the objects in S, a second matrix, $\mathbf{V} = \{v_{ij}\}$, is also given having dichotomous (0/1) entries. Here, $v_{ij} = 0$ if p_{ij} is *not* to be considered (i.e., the proximity value is missing) in the calculation of whatever merit or cost measure is being optimized, and $= 1$ otherwise. For example, in the partitioning task, if the average within subset proximity were used as a heterogeneity measure for a subset, say, S_a, it could be defined as

$$\left(\frac{1}{\sum_{i,j\in S_a} v_{ij}}\right) \sum_{i,j\in S_a} v_{ij} p_{ij},$$

i.e., the average of all the within subset proximities in S_a that are not missing. Analogously, in a sequencing context and for a comparison using a gradient measure, say, among three objects $O_i \prec O_j \prec O_k$, the comparison between

p_{ij} and p_{ik} would be ignored if either $v_{ij} = 0$ or $v_{ik} = 0$, and not included in the calculation of the overall merit measure. If we wish, a set of weights for objects (indicating replication) and a set of weights identifying those proximities that are missing could be combined into a single matrix, say, $\{v_{ij}w_iw_j\}$. Here, $v_{ij}w_iw_j = 0$ if the proximity between the object (types) O_i and O_j is missing and $= w_iw_j$ otherwise, indicating the number of replications that particular proximity value should have in the calculation of whatever cost or merit measure is being optimized.

There is one specific pattern of missing proximities that we have already discussed implicitly in previous chapters whenever two-mode proximity matrices were considered. In these instances, two disjoint sets, S_A and S_B, were given and the proximity matrix $\mathbf{P}^{AB}$ was defined on the combined set, $S = S_A \cup S_B$, but with missing proximities within S_A and within S_B. This two-mode usage can be rephrased through an indicator matrix $\{v_{ij}\}$, where $v_{ij} = 0$ if $O_i, O_j \in S_A$ or $O_i, O_j \in S_B$, and $= 1$ otherwise. More generally, however, the incorporation of other patternings of missing proximities is possible for all the specializations of the GDPP considered, and most (but not all) measures of cost or merit could be generalized to accommodate them. There are some exceptions, as in the use of coordinate estimation for sequencing within a GDPP framework, where such an accommodation is not possible. This fact was observed for the very regular patterning of missing proximities in our discussion in Section 4.1.3 on the difficulties of incorporating two-mode proximity data within a DP framework when coordinate estimation was the means for constructing the measure of merit to be maximized.

6.2 Prospects

This monograph has attempted, by example, to show the wide possibilities inherent in the general framework of dynamic programming (and for the general dynamic programming paradigm [GDPP] in particular) for approaching a variety of combinatorial data analysis (CDA) tasks. At one level, this extended presentation gives a convenient archival source of reference for the range of nontrivially-sized CDA tasks that can be handled by the GDPP, and all with a guarantee of generating optimal solutions. At a different, more applied level, and because our primary objective is to be didactic in our presentation of the uses of the GDPP, we have given the individual interested in trying out these methods a mechanism for doing so, and in turn, for verifying how the methods work as advertised. The Appendix presents how the various program implementations may be obtained and, by following the documentation presented, how they can be used. We know of no other source (commercial or otherwise) that provides such an already implemented computational possibility.

The GDPP is a very general strategy for approaching a variety of CDA tasks, and because of this generality, there will be instances in which better optimization methods for specific problems can be defined. Some of these cases have been mentioned in the course of our presentation (e.g., in the linear assignment task and the construction of optimal paths, among others). In those cases, it is still relevant that we can construct benchmarks (generated from the GDPP) that other optimization methods proposed for specific CDA tasks can seek to better. For instances where the domain of search can be appropriately restricted, and large-sized optimization tasks handled optimally by the GDPP, it may even be that the GDPP is the optimization method of choice (e.g., in partitioning tasks subject to order constraints on the defining clusters).

One aspect of the use of the GDPP introduced in Chapter 5 that seems very promising is the adoption of the recursive principles heuristically for approaching (very) large object sets. This usage obviously needs more evaluative study in the future, but preliminary results suggest that such a heuristic strategy relying on large "chunks" of the object set of interest (but with a way of investigating and improving upon the formation of the "chunks") may be much better than the more typical local operation methods currently in widespread use. There is also much more to be done in implementing the GDPP in the most efficient ways possible for specific tasks at hand (some of this work is ongoing at present; for instance, see endnote 45 to the appendix). In short, the prospects appear to be excellent for much further work in the CDA area that builds on the presentation given in this monograph. We hope that readers, who have come this far, agree.

Endnote

[44] Although we will not pursue the possibility in any great detail, we might comment that the observation just made about $\mathcal{F}(A_k)$ providing the optimal contribution to a total merit measure if the objects in A_k either appear in the *first* k or the *last* k positions suggests a mechanism for reducing the computational storage requirements for finding an optimal ordering in the first place. Specifically, instead of defining a recursive process over a collection of n sets $\Omega_1, \ldots, \Omega_n$, where Ω_k contains all k-element subsets of S, it would be sufficient to proceed with the recursion from Ω_1 only to $\Omega_{[(n+1)/2]}$, where $[x]$ denotes the greatest integer contained in x. Once completed, an optimal order could be identified directly from the maximum of the sums of $\mathcal{F}(A_{[(n+1)/2]})$ for $A_{[(n+1)/2]} \in \Omega_{[(n+1)/2]}$ and $\mathcal{F}(S - A_{[(n+1)/2]})$ for $S - A_{[(n+1)/2]} \in \Omega_{n-[(n+1)/2]}$. This process of 'doubling-up' computationally (i.e., proceeding with the recursion only partially through a collection of sets $\Omega_1, \ldots, \Omega_K$, and then identifying an optimal solution though a search process on the information contained within this partial collection) has analogues for the construction of optimal partitions as well, whether ordered or not.

Appendix

Available Programs

The purpose of this appendix is to describe briefly the operation of the nineteen programs mentioned in the endnotes throughout the monograph. A summary list is first given below that groups these nineteen programs into five general task areas: (1) object partitioning, (2) hierarchical clustering, (3) object sequencing/seriation, (4) object sequencing by constructing optimal paths, and (5) constructing ordered partitions. The various program acronyms are given, along with an indication of how the acronyms were constructed. Also, a brief statement is provided as to what each program is intended to do. These short summaries are elaborated upon in the five main sections of this appendix proper, which discuss the operation of each program under its general task area. The various options/alternatives offered are noted, and sample input/output provided.

Summary List of Available Programs

(1) Object partitioning—

DPCL1U (Dynamic Programming CLustering 1-mode Unrestricted): the primary partitioning program for a symmetric proximity matrix. There are twelve options for defining subset heterogeneity, and two optimization criteria of minimizing either the sum or the maximum heterogeneity measure over the classes of a partition.

DPCL1R (Dynamic Programming CLustering 1-mode Restricted): all the same options as DPCL1U for a symmetric proximity matrix, but with the additional requirement that an object order be given; all subsets considered contain objects consecutive with respect to this ordering.

DPCL2U (Dynamic Programming CLustering 2-mode Unrestricted): the analogue of DPCL1U but now for two-mode data that are provided in the form of a rectangular proximity matrix between two disjoint sets.

DPCL2R (Dynamic Programming CLustering 2-mode Restricted): all the same options as DPCL1R for a two-mode data matrix, but with the additional requirement that separate row and column object orders be given; all subsets considered include row and column objects that are consecutive with respect to these two orderings.

HPCL1U (Heuristic Programming CLustering 1-mode Unrestricted): the extension of DPCL1U for symmetric proximity matrices to allow (a) the entities to be partitioned to be subsets of the original object set, and (b) only parts of the original object set to be partitioned.

HPCL2U (Heuristic Programming CLustering 2-mode Unrestricted): the analogue of DPCL2U with all its same options for two-mode proximity matrices and allowing (a) the entities to be partitioned to be subsets of the combined row and column object set, and (b) only parts of the combined row and column object set to be partitioned.

(2) Hierarchical clustering—

DPHI1U (Dynamic Programming HIerarchical clustering 1-mode Unrestricted): the primary hierarchical clustering program for a symmetric proximity matrix. There are fourteen options for defining subset heterogeneity, with twelve identical to DPCL1U in the partitioning context and two devoted to the least-squares fitting of an ultrametric and the two admissibility criteria discussed in the text. The sum of the heterogeneity measures is minimized over the new classes formed; when appropriate, an additional partition hierarchy based on the 'greedy' strategy is also given (which provides a computational bound in obtaining the optimal hierarchy).

DPHI1R (Dynamic Programming HIerarchical clustering 1-mode Restricted): all the same options as DPHI1U for a symmetric proximity matrix, but with the additional requirement that an object order be given; all new subsets formed in the process of partition hierarchy construction contain objects consecutive with respect to this ordering.

DPDI1U (Dynamic Programming DIvisive clustering 1-mode Unrestricted): all the same options as DPHI1U for a symmetric proximity matrix, but the algorithm proceeds divisively. Thus, in using DPDI1, the optimal partial partition hierarchies provided all begin with the all-inclusive set; in contrast, for DPHI1U the optimal partial partition hierarchies all begin with each object forming its own separate class.

HPHI1U (Heuristic Programming HIerarchical clustering 1-mode Unrestricted): the extension of DPHI1U for symmetric proximity matrices, that allows (a) the entities to be partitioned to be subsets of the original object set, and (b) only parts of the original object set to be partitioned. The L_1 and L_∞ options are included, along with the construction of hierarchies based only on the order properties of the proximities (i.e., using OTIs and OQIs).

HPHI2U (Heuristic Programming HIerarchical clustering 2-mode Unrestricted): the analogue of HPHI1U but now for two-mode data that are provided in the form of a rectangular proximity matrix between two disjoint sets. Also, (a) the entities to be partitioned can be subsets of the original object set, and (b) only parts of the original object set need be partitioned.

APPENDIX: AVAILABLE PROGRAMS

(3) Object sequencing/seriation—

DPSE1U (<u>D</u>ynamic <u>P</u>rogramming <u>SE</u>riation <u>1</u>-mode <u>U</u>nrestricted): the primary sequencing/seriation program for a symmetric or skew-symmetric proximity matrix. There are eight options for defining the index of matrix pattern, depending on whether the proximity matrix is symmetric or skew-symmetric. In all cases, the index of matrix pattern is optimized over all possible reorderings of the rows/columns of the proximity matrix. Precedence constraints on the ordering can be imposed through an additional indicator matrix that is read after the proximity matrix and is present in the same input file. Also, the option of using only the within row or within column gradient comparisons is provided when these indices of matrix pattern are chosen.

HPSE1U (<u>H</u>euristic <u>P</u>rogramming <u>SE</u>riation <u>1</u>-mode <u>U</u>nrestricted): the extension of DPSE1U for symmetric or skew-symmetric proximity matrices that allows (a) the entities to be sequenced to be subsets of the original object set, and (b) only parts of the original object set to be sequenced.

DPSE2U (<u>D</u>ynamic <u>P</u>rogramming <u>SE</u>riation <u>2</u>-mode <u>U</u>nrestricted): the primary sequencing/seriation program for a two-mode proximity matrix, where the row and column objects are treated as one combined set. There are four options for defining the index of matrix pattern—the weighted and unweighted gradient measures (to be maximized), plus the two that use only the weighted or unweighted inconsistencies (which are minimized). The comparisons can be restricted to be within only the rows or columns of the proximity matrix for the combined object set, or only within the rows or the columns of the original two-mode proximity matrix. Also, precedence constraints on the combined row and column object set can be imposed through an additional indicator matrix read in after the two-mode proximity matrix.

HPSE2U (<u>H</u>euristic <u>P</u>rogramming <u>SE</u>riation <u>2</u>-mode <u>U</u>nrestricted): the extension of DPSE2U for two-mode proximity matrices that allows (a) the entities to be sequenced to be subsets of the original combined row/column object set, and (b) only parts of the combined row/column object set to be sequenced.

DPSE2R (<u>D</u>ynamic <u>P</u>rogramming <u>SE</u>riation <u>2</u>-mode <u>R</u>estricted): all the same options as DPSE2U for a two-mode data matrix, but allowing object orders to be given for the column objects only, or for both the row and column objects. When both row and column orders are imposed, the DHG measure of matrix pattern is an additional option.

(4) Object sequencing by constructing optimal paths—

DPSEPH (<u>D</u>ynamic <u>P</u>rogramming <u>SE</u>riation by constructing <u>P</u>at<u>H</u>s): the sole program for constructing either linear or circular paths based on a given one-mode proximity matrix. The optimization task may be to find the minimum or maximum path, or one that is minimax or maximin. Precedence constraints can be imposed by the use of an indicator matrix read in after the proximity matrix. If the proximity matrix is square but nonsymmetric, directed linear or circular paths are identified that satisfy the optimization criteria chosen.

(5) Constructing ordered partitions—

DPOP1U (Dynamic Programming Ordered Partitioning 1-mode Unrestricted): for a symmetric or skew-symmetric proximity matrix, optimal ordered partitions are identified using one of five different options (three for symmetric matrices and two for skew-symmetric matrices).

HPOP1U (Heuristic Programming Ordered Partitioning 1-mode Unrestricted): the extension of DPOP1U for symmetric and skew-symmetric proximity matrices to allow (a) the entities to be partitioned into ordered classes to be subsets of the original object set, and (b) only parts of the original object set to be partitioned.

The nineteen programs summarized above are available in executable forms (i.e., as *.exe files) that will run under 32 bit operating systems such as Windows 95 or Windows NT. These programs can be downloaded from the following WWW site:

ftp://www.psych.uiuc.edu/pub/lhubert

There are four files maintained at that address:

PROG32.ZIP, which includes the nineteen *.exe files in 'zipped' form;

CODE32.ZIP, which includes the original Fortran90 source code for each of the nineteen programs in 'zipped' form;

DATA.ZIP, which includes the data files that have been used throughout the monograph and in this appendix for all the various examples;

PKUNZIP.EXE, which will unzip the first three files.

Downloading all four files into a directory and then issuing the commands of 'PKUNZIP' plus the file name will provide the executable files for all nineteen programs, along with the various data sets that could be used to reproduce the analyses given throughout the monograph, and the original Fortran90 source code for the nineteen programs that were compiled to produce the *.exe files. By default, all programs have been compiled under Microsoft Fortran Powerstation 4.0 and, thus, require execution under a 32 bit operating system such as Windows 95 or Windows NT.[45]

Object Partitioning

The primary program for the partitioning task is DPCL1U; the five other programs that are devoted to partitioning (DPCL1R, DPCL2R, DPCL2U, HPCL1U, HPCL2U) can be viewed as variants of DPCL1U in that they handle two-mode proximity data, impose order restrictions, or allow heuristic extensions for larger object sets. We will start with DPCL1U and go through the steps for one representative application and include a listing of the output produced. Once this example is given, the variations present within the other five programs in the series will be noted, but representative applications will not be given explicitly.

To begin, we will assume that the directory from which DPCL1U is run has a file containing a proximity matrix in lower-triangular form without main

OBJECT PARTITIONING

diagonal entries. The example we use is from the digit data in Table 1.1, where the file is named 'number.dat', and has the (fixed) format (in the usual Fortran form) for each of its lines of '(9F4.3)':

```
421
584 284
709 346 354
684 646 059 413
804 588 671 429 409
788 758 421 300 388 396
909 630 796 592 742 400 417
821 791 367 804 246 671 350 400
850 625 808 263 683 592 296 459 392
```

For an illustration, suppose we select as a heterogeneity measure the average proximity within a subset (based on the number of object pairs), the optimization criterion of minimizing the maximum heterogeneity measure over all subsets, and then construct optimal partitions into 2 through 10 subsets (so the maximum number of subsets selected will be 10, which will automatically generate all optimal partitions into 2 through 10 subsets as well).

Running DPCL1U produces the prompts given in all capital letters below. The responses provided by the user to effect the desired analysis are placed in solid braces [·] (but note that these solid braces should not be part of the input), and explanatory comments about what has been chosen and why are given in curly braces.

```
MAIN MENU
1   AVERAGE PROXIMITY WITHIN SUBSET (USING TWICE
    THE NUMBER OF OBJECTS)
2   AVERAGE PROXIMITY WITHIN SUBSET (USING THE
    NUMBER OF OBJECT PAIRS)
3   SUM OF PROXIMITIES WITHIN SUBSET
4   MAXIMUM PROXIMITY WITH SUBSET
5   AVERAGE WITHIN MINUS AVERAGE BETWEEN
6   NUMBER OF PROXIMITIES WITHIN GREATER THAN BETWEEN
7   NORMALIZED OPTION 6
8   COMMON OBJECT INCONSISTENCIES WITHIN VERSUS
    BETWEEN
9   NORMALIZED OPTION 8
10  NEGATIVE OF THE AVERAGE BETWEEN
11  MINIMUM CONNECTIVITY PROXIMITY
12  SPANNING TREE LENGTH
CHOICE?
[2]
```

{Each of the twelve subset heterogeneity options from Section 3.1 is represented in this list; the Roman numerals used in Section 3.1 correspond to these numbered options as follows: 1:(iii); 2:(ii); 3:(i); 4:(iv); 5:(vi); 6:(viii);

7:(ix); 8:(x); 9:(xi); 10:(vii); 11(v); 12:(xii). Because we wish to use the average within-subset proximity based on the number of object pairs, option 2 is chosen.}

```
OPTIMIZATION METHOD
 1 MINIMUM OF THE SUM
 2 MINIMUM OF THE MAXIMUM
CHOICE?
[2]
```

{We wish to minimize the maximum heterogeneity measure over all subsets, so option 2 is chosen.}

```
MAXIMUM NUMBER OF SUBSETS TO BE USED
[10]
```

{All optimal partitions into 2 through 10 subsets will be generated.}

```
NUMBER OF ROWS/COLUMNS FOR THE LOWER TRIANGULAR
MATRIX-- NO DIAGONAL
[10]
```

{Because the number of objects to be partitioned is 10, this number is provided as a response. If a number is given that is less than 3 or greater than 30, an error message appears, declaring that the 'PROBLEM SIZE IS OUT OF RANGE', and the program stops. Otherwise, the program tries to allocate the necessary storage space required, and if enough RAM exists, will respond with 'ARRAY ALLOCATION SUCCESSFUL' (twice); if not, the response will be 'ARRAY ALLOCATION ERROR', and the program terminates.}

```
FILE NAME FOR THE INPUT MATRIX
[number.dat]
```

{This file, containing the lower-triangular proximity matrix, must exist in the directory, otherwise an error message will appear after the format is input below. (Alternatively, a path could be given for where the input file can be found.)}

```
FILE NAME FOR THE OUTPUT MATRIX
[number.out]
```

{The output will appear in a new file named 'number.out'; if a file by this name already exists, an error message will be given after the format is input below. (Alternatively, a path could be given for where the output file should be written.)}

PROVIDE FORMAT FOR THE INPUT MATRIX
[(9F4.3)]

{The appropriate Fortran format statement is given for a line of the proximity matrix.}

PROGRAM RUNNING
STAGE ONE PROCESSING SUBSETS OF SIZE 1

⋮

STAGE ONE PROCESSING SUBSETS OF SIZE 10
PROCESSING A STAGE   2

⋮

PROCESSING A STAGE   10

{No response is required; these are merely messages that indicate the processing has begun and that it is completed when the phrase 'PROCESSING A STAGE 10' appears.}

The contents of the file 'number.out' are given below (with some intermediate output deleted), where the various items should be self-explanatory. There is first a listing of which subset heterogeneity option and optimization criterion were used, and a listing of the proximity matrix. Then, for partitions into 1 to 10 subsets, an indication is given of the heterogeneity indices of the classes within each of the optimal partitions, and a cumulative index over all classes of that partition.

```
HETEROGENEITY OPTION    2
MINIMIZATION OF THE MAXIMUM
INPUT MATRIX
    0.421
    0.584   0.284
    0.709   0.346   0.354
    0.684   0.646   0.059   0.413
    0.804   0.588   0.671   0.429   0.409
    0.788   0.758   0.421   0.300   0.388   0.396
    0.909   0.630   0.796   0.592   0.742   0.400   0.417
    0.821   0.791   0.367   0.804   0.246   0.671   0.350   0.400
    0.850   0.625   0.808   0.263   0.683   0.592   0.296   0.459   0.392
```

```
NUMBER OF SUBSETS    1

SUBSET ADDED 10  9  8  7  6  5  4  3  2  1
INDEX FOR SUBSET         0.5412   CUMULATIVE INDEX      0.5412


NUMBER OF SUBSETS    2

SUBSET ADDED 10  9  8  7  6
INDEX FOR SUBSET         0.4373   CUMULATIVE INDEX      0.4500

SUBSET ADDED  5  4  3  2  1
INDEX FOR SUBSET         0.4500   CUMULATIVE INDEX      0.4500


NUMBER OF SUBSETS    3

SUBSET ADDED 10  8  7  6  4
INDEX FOR SUBSET         0.4144   CUMULATIVE INDEX      0.4144

SUBSET ADDED  9  5  3  2
INDEX FOR SUBSET         0.3988   CUMULATIVE INDEX      0.3988

SUBSET ADDED  1
INDEX FOR SUBSET         0.0000   CUMULATIVE INDEX      0.0000


NUMBER OF SUBSETS    4

SUBSET ADDED 10  9  8  7
INDEX FOR SUBSET         0.3857   CUMULATIVE INDEX      0.3857

SUBSET ADDED  6
INDEX FOR SUBSET         0.0000   CUMULATIVE INDEX      0.3503

SUBSET ADDED  5  4  3  2
INDEX FOR SUBSET         0.3503   CUMULATIVE INDEX      0.3503

SUBSET ADDED  1
INDEX FOR SUBSET         0.0000   CUMULATIVE INDEX      0.0000

NUMBER OF SUBSETS    5

SUBSET ADDED  8
INDEX FOR SUBSET         0.0000   CUMULATIVE INDEX      0.3503

SUBSET ADDED 10  9  7
INDEX FOR SUBSET         0.3460   CUMULATIVE INDEX      0.3503

SUBSET ADDED  6
INDEX FOR SUBSET         0.0000   CUMULATIVE INDEX      0.3503

SUBSET ADDED  5  4  3  2
INDEX FOR SUBSET         0.3503   CUMULATIVE INDEX      0.3503

SUBSET ADDED  1
INDEX FOR SUBSET         0.0000   CUMULATIVE INDEX      0.0000
```

{The remaining output for 6 to 10 subsets has not been printed.}

Program DPCL1R:

The program DPCL1R is an analogue of DPCL1U in that the same options are offered, but it requires the additional imposition of an order constraint on which objects can be defined within each subset of an optimal partition. (Also, much larger proximity matrices are allowable than in DPCL1U.) After the input of the format, we have:

```
0 IF CONSTRAINED ORDER IS TO BE THE IDENTITY
1 IF IT IS TO BE INPUT
[1]
```

{A '0' response imposes the order constraint that is the same as the index order of the objects; a '1' allows the order constraint to be supplied (which we do in this example).}

```
ENTER ORDER SEPARATED BY BLANKS
[1 2 3 4 5 6 7 8 9 10]
```

{The object order used here again provides the (identity) index order for the objects (and could also have been indicated by the use of a '0' in the previous response).}

The output is the same as that for DPCL1U except that the subsets in any optimal partition would now contain objects consecutive in the object order imposed. For reference, a listing of the object ordering that is used for the constraint is given in the output file immediately before the input matrix is printed.

Program DPCL2U:

The program DPCL2U constructs optimal partitions based on two-mode data and offers all the same options as DPCL1U. Thus, the data matrix to be input must be in a rectangular form, and the format statement applies to each of its rows. The prompt for the 'NUMBER OF OBJECTS FOR A LOWER TRIANGULAR MATRIX– NO DIAGONAL' is replaced by two separate prompts:

```
NUMBER OF ROWS
[?]
NUMBER OF COLUMNS
[?]
```

where '?' is replaced by the appropriate number. The prompt for the 'MAXIMUM NUMBER OF SUBSETS TO BE USED' is augmented by a reminder:

```
NO LARGER THAN THE MINIMUM OF THE NUMBER OF
ROWS AND COLUMNS
[?]
```

After the number or rows and columns are input, if this condition does not hold, the program terminates with an error message.

In the output, the data matrix echoed is now rectangular; in the presentation of the clustering results, the row and column objects are treated as one joint set with the column objects labeled by the appropriate column index plus the number of row objects (e.g., the first column object has a label of 1 + number of row objects). A reminder of this particular column object labeling convention is also given in the output.

Program DPCL2R:

The program DPCL2R is the analogue of DPCL1R but requires the imposition of separate orders for the row objects and for the column objects. The prompts that ask for these have the form

```
0 IF ROW CONSTRAINED ORDER IS TO BE THE IDENTITY
1 IF IT IS TO BE INPUT
[1]

ENTER ROW ORDER SEPARATED BY BLANKS
[...row order here...]

0 IF COLUMN CONSTRAINED ORDER IS TO BE THE IDENTITY
1 IF IT IS TO BE INPUT
[1]

ENTER COLUMN ORDER SEPARATED BY BLANKS
[...column order here...]
```

In the output file these two row and column constraint orders are provided before the input matrix; in the listing of the subsets of an optimal partition, the row objects and the column objects making up a subset are given separately.

Program HPCL1U:

Program HPCL1U is the extension of DPCL1U that allows user-specified classes of objects to form the entities to be partitioned. After the request for the 'NUMBER OF ROWS/COLUMNS FOR THE LOWER TRIANGULAR MATRIX', the following prompt is given:

```
NUMBER OF INITIAL OBJECT CLASSES TO BE PARTITIONED?
[?]
```

This question refers to the number of object classes that will be defined (in response to a later prompt) and treated as the entities to be partitioned. The next prompt of

```
MAXIMUM NUMBER OF SUBSETS TO BE USED
[?]
```

now refers to the maximum number of subsets desired in an optimal partition of the (to be defined below) entities to be partitioned. They are constructed according to the response to

```
OBJECT CLASSES TO BE PARTITIONED?
1 THE SAME AS THE NUMBER OF OBJECTS
2 OBJECT CLASS MEMBERSHIP TO BE INPUT
[2]
```

A response of '1' would define as many object classes as there are original objects with the same integer labels (i.e., DPCL1U is obtained as a special case); a response of '2' allows the class membership to be constructed by producing the prompt

```
INPUT CLASS MEMBERSHIP (SEPARATED BY BLANKS)
WITH ZERO INDICATING A DELETED OBJECT
[...integer class membership labels...]
```

The class membership labels use the integers from 1 to the number of entities defined, or zero; all the initial objects that are given a common integer label provide a class that is treated as a single entity in the partitioning process; if an initial object is given a '0', it appears in no such entity class. In the output, a legend giving the 'CLASS NUMBER / OBJECT MEMBERSHIP IN CLASS' is provided; all the results are given using the class numbers from this legend.

Program HPCL2U:

Program HPCL2U is an extension of DPCL1U for a two-mode proximity matrix and allows user-specified classes of objects to form the entities to be partitioned. In comparison with the earlier discussion of DPCL1U and HPCL1U, the only prompt requiring further clarification is the one that defines the group membership:

```
INPUT CLASS MEMBERSHIP (SEPARATED BY BLANKS)
WITH ZEROS INDICATING A DELETED OBJECT
THE ROW OBJECTS COME FIRST AND THEN THE COLUMN
[...integer class membership labels for rows...integer
class membership labels for columns...]
```

The original object set is a joint set of the row and column objects, and the class membership labels for the rows should be given first followed by the class membership labels for the column objects (with a '0' listed whenever an object is to be deleted). Thus, the entities to be defined and partitioned could consist of all row or all column objects, or be a mixture of the two. Again, as in the output for HPCL1U, a legend indicating 'CLASS NUMBER / OBJECT MEMBERSHIP IN CLASS' is provided, and all the results are given using the class numbers from this legend.

Hierarchical Clustering

Analogous to the software for object partitioning, there is one primary program, DPHI1U, for the hierarchical clustering task with four others (DPHI1R, DPDI1U, HPHI1U, HPHI2U) as variants dealing with order restrictions on the objects, a divisive approach to constructing the optimal hierarchy, and heuristic extensions for larger object sets for both one- and two-mode proximity matrices. We will illustrate explicitly the use of DPHI1U on the digit data of Table 1.1 (using the input file 'number.dat'), and choose the average proximity within a subset (based on the number of object pairs) to construct the optimal partition hierarchy that minimizes the sum of these averages over all the new subsets that are formed in a hierarchy. A copy of the output produced (in the file 'numhi.out') will be included below. After this example, the variations provided by the other four programs in this series will be noted, but no specific applications will be given for them.

Running DPHI1U produces the following prompts in capital letters (we again give the responses provided by the user in solid braces [·] and provide explanatory comments when necessary in curly braces).

```
MAIN MENU
1  AVERAGE PROXIMITY WITHIN SUBSET (USING TWICE
   THE NUMBER OF OBJECTS)
2  AVERAGE PROXIMITY WITHIN SUBSET (USING THE
   NUMBER OF OBJECT PAIRS)
3  SUM OF PROXIMITIES WITHIN SUBSET
4  MAXIMUM PROXIMITY WITH SUBSET
5  AVERAGE WITHIN MINUS AVERAGE BETWEEN
6  NUMBER OF PROXIMITIES WITHIN GREATER THAN BETWEEN
7  NORMALIZED OPTION 6
8  COMMON OBJECT INCONSISTENCIES WITHIN VERSUS
   BETWEEN
9  NORMALIZED OPTION 8
10 NEGATIVE OF THE AVERAGE BETWEEN
11 MINIMUM CONNECTIVITY PROXIMITY
12 SPANNING TREE LENGTH
13 BETWEEN SUM OF SQUARES FROM MEAN WITH ORDER
   CONSTRAINTS PLUS LATER UNION CONSTRAINTS
14 BETWEEN SUM OF SQUARES FROM MEAN WITH ORDER
   CONSTRAINTS ONLY
```

HIERARCHICAL CLUSTERING

CHOICE?
[2]

{In addition to the same options for the subset heterogeneity measure numbered 1 to 12, as for the partitioning task, the two options of 13 and 14 (as discussed in the text) provide the least-squares fitting of an ultrametric using either the possibly too lenient admissibility criterion (option 14) or the possibly too strict admissibility criterion (option 13).}

IMPOSITION OF NEW SUBSET ORDER CONSTRAINTS
 1 NO
 2 YES
CHOICE?
[2]

{In options 1, 2, 3, 4, 11, and 12, where in an optimal solution the subset heterogeneity measures must be monotonic, such a requirement can be imposed in constructing the partition hierarchy so as to reduce the overall computational effort required. In addition, for options other than 5, 10, 13, or 14, a greedy partition is first constructed that provides a computational upper bound (as noted in the text) on the value that an optimal hierarchy should achieve; this bound is also used to reduce the computational effort in constructing the optimal hierarchy and is indicated explicitly in the output as well.}

NUMBER OF ROWS/COLUMNS FOR THE LOWER TRIANGULAR
MATRIX-- NO DIAGONAL
[10]

{If a number is entered that is less than or equal to 3 or greater than 15, an error message appears in the form 'MATRIX OUT OF RANGE' and the program stops. Otherwise, the program attempts to allocate the necessary storage space, and if successful, the message 'ARRAY ALLOCATION SUCCESSFUL' will appear (three times); if not, a message of 'ARRAY ALLOCATION ERROR' occurs, and the program terminates.}

FILE NAME FOR THE INPUT MATRIX
[number.dat]

FILE NAME FOR THE OUTPUT MATRIX
[numhi.out]

PROVIDE FORMAT FOR THE INPUT MATRIX
[(9F4.3)]

PROGRAM RUNNING

{No additional prompts are given until the processing is completed.}

The contents of the output file 'numhi.out' are reproduced below, where the various entries should be more or less self-explanatory. There are indications of which subset heterogeneity measure (transition criterion) is being used; that we are maximizing the sum of these in constructing the partition hierarchy; that subset order constraints are being imposed; that a greedy algorithm is used to obtain a computational bound; and that a listing is given of the input proximity matrix. Thereafter, the greedy partition hierarchy is provided along with the bound used, the cumulative index up to the given number of subsets in the hierarchy, and a listing of group membership for these subsets. Next, the optimal hierarchy is given plus the optimal partial hierarchies that stop at 2 through 9 subsets. If either of the least-squares options 13 or 14 had been used, the average proximity between each pair of subsets at a given level is also provided directly before the listing of subset membership.

```
TRANSITION CRITERION USED    2

MINIMIZATION OF THE SUM

SUBSET ORDER CONSTRAINTS IMPOSED

PRELIMINARY GREEDY ALGORITHM USED FOR A BOUND

INPUT MATRIX

    0.421

    0.584    0.284

    0.709    0.346    0.354

    0.684    0.646    0.059    0.413

    0.804    0.588    0.671    0.429    0.409

    0.788    0.758    0.421    0.300    0.388    0.396

    0.909    0.630    0.796    0.592    0.742    0.400    0.417

    0.821    0.791    0.367    0.804    0.246    0.671    0.350    0.400

    0.850    0.625    0.808    0.263    0.683    0.592    0.296    0.459    0.392


BOUND USED         2.515200
GREEDY PARTITION HIERARCHY TO OBTAIN THE BOUND
SUBSETS    1  INDEX       3.0564 MEMBERSHIP   1  1  1  1  1  1  1  1  1  1
SUBSETS    2  INDEX       2.5152 MEMBERSHIP   1  1  1  2  1  2  2  2  1  2
SUBSETS    3  INDEX       2.0249 MEMBERSHIP   1  2  2  3  2  3  3  3  2  3
SUBSETS    4  INDEX       1.6105 MEMBERSHIP   1  2  2  3  2  3  3  4  2  3
SUBSETS    5  INDEX       1.2117 MEMBERSHIP   1  2  3  4  3  4  4  5  3  4
SUBSETS    6  INDEX       0.8323 MEMBERSHIP   1  2  3  4  3  5  4  6  3  4
SUBSETS    7  INDEX       0.5460 MEMBERSHIP   1  2  3  4  3  5  6  7  3  4
SUBSETS    8  INDEX       0.2830 MEMBERSHIP   1  2  3  4  3  5  6  7  3  8
SUBSETS    9  INDEX       0.0590 MEMBERSHIP   1  2  3  4  3  5  6  7  8  9
SUBSETS   10  INDEX       0.0000 MEMBERSHIP   1  2  3  4  5  6  7  8  9 10


OPTIMAL PARTITION HIERARCHY
SUBSETS    1  INDEX       3.0547 MEMBERSHIP   1  1  1  1  1  1  1  1  1  1
```

HIERARCHICAL CLUSTERING

```
SUBSETS   2  INDEX    2.5135 MEMBERSHIP  1 1 2 2 2 2 2 2 2
SUBSETS   3  INDEX    2.0450 MEMBERSHIP  1 1 2 3 2 2 3 3 2 3
SUBSETS   4  INDEX    1.6240 MEMBERSHIP  1 2 3 4 3 3 4 4 3 4
SUBSETS   5  INDEX    1.2202 MEMBERSHIP  1 2 3 4 3 5 4 4 3 4
SUBSETS   6  INDEX    0.8323 MEMBERSHIP  1 2 3 4 3 5 4 6 3 4
SUBSETS   7  INDEX    0.5460 MEMBERSHIP  1 2 3 4 3 5 6 7 3 4
SUBSETS   8  INDEX    0.2830 MEMBERSHIP  1 2 3 4 3 5 6 7 3 8
SUBSETS   9  INDEX    0.0590 MEMBERSHIP  1 2 3 4 3 5 6 7 8 9
SUBSETS  10  INDEX    0.0000 MEMBERSHIP  1 2 3 4 5 6 7 8 9 10


OPTIMAL PARTITION HIERARCHY STARTING WITH  2 SUBSETS
SUBSETS   2  INDEX    2.5135 MEMBERSHIP  1 1 2 2 2 2 2 2 2
SUBSETS   3  INDEX    2.0450 MEMBERSHIP  1 1 2 3 2 2 3 3 2 3
SUBSETS   4  INDEX    1.6240 MEMBERSHIP  1 2 3 4 3 3 4 4 3 4
SUBSETS   5  INDEX    1.2202 MEMBERSHIP  1 2 3 4 3 5 4 4 3 4
SUBSETS   6  INDEX    0.8323 MEMBERSHIP  1 2 3 4 3 5 4 6 3 4
SUBSETS   7  INDEX    0.5460 MEMBERSHIP  1 2 3 4 3 5 6 7 3 4
SUBSETS   8  INDEX    0.2830 MEMBERSHIP  1 2 3 4 3 5 6 7 3 8
SUBSETS   9  INDEX    0.0590 MEMBERSHIP  1 2 3 4 3 5 6 7 8 9
SUBSETS  10  INDEX    0.0000 MEMBERSHIP  1 2 3 4 5 6 7 8 9 10

OPTIMAL PARTITION HIERARCHY STARTING WITH  3 SUBSETS
SUBSETS   3  INDEX    2.0249 MEMBERSHIP  1 2 2 3 2 3 3 3 2 3
SUBSETS   4  INDEX    1.6105 MEMBERSHIP  1 2 2 3 2 3 3 4 2 3
SUBSETS   5  INDEX    1.2117 MEMBERSHIP  1 2 3 4 4 4 5 3 4
SUBSETS   6  INDEX    0.8323 MEMBERSHIP  1 2 3 4 3 5 4 6 3 4
SUBSETS   7  INDEX    0.5460 MEMBERSHIP  1 2 3 4 3 5 6 7 3 4
SUBSETS   8  INDEX    0.2830 MEMBERSHIP  1 2 3 4 3 5 6 7 3 8
SUBSETS   9  INDEX    0.0590 MEMBERSHIP  1 2 3 4 3 5 6 7 8 9
SUBSETS  10  INDEX    0.0000 MEMBERSHIP  1 2 3 4 5 6 7 8 9 10

OPTIMAL PARTITION HIERARCHY STARTING WITH  4 SUBSETS
SUBSETS   4  INDEX    1.6105 MEMBERSHIP  1 2 2 3 2 3 3 4 2 3
SUBSETS   5  INDEX    1.2117 MEMBERSHIP  1 2 3 4 3 4 4 5 3 4
SUBSETS   6  INDEX    0.8323 MEMBERSHIP  1 2 3 4 3 5 4 6 3 4
SUBSETS   7  INDEX    0.5460 MEMBERSHIP  1 2 3 4 3 5 6 7 3 4
SUBSETS   8  INDEX    0.2830 MEMBERSHIP  1 2 3 4 3 5 6 7 3 8
SUBSETS   9  INDEX    0.0590 MEMBERSHIP  1 2 3 4 3 5 6 7 8 9
SUBSETS  10  INDEX    0.0000 MEMBERSHIP  1 2 3 4 5 6 7 8 9 10
```

{The remaining output for hierarchies starting at 5 to 9 subsets has not been printed.}

Program DPHI1R:

The program DPHI1R is a direct analogue of DPHI1U in offering the same options, but requires the additional imposition of an order constraint on which objects can be within a subset of a partition in the hierarchy. The prompt is issued and responded to exactly as in DPCL1R. (If the number of objects entered is less than or equal to 3 or greater than or equal to 30, an error message is generated that the 'MATRIX IS OUT OF RANGE'; otherwise, the program will attempt to allocate the storage space necessary to construct an optimal partition hierarchy.)

Program DPDI1U:

The only difference between DPDI1U and DPHI1U is that the former constructs an optimal partition hierarchy divisively rather than agglomeratively. The prompts in running both programs are identical. In the output for DPDI1U, however, the optimal partial partition hierarchies that are provided begin with the all-inclusive set and terminate at 2 through 9 subsets (if the given number of the initial objects is 10, as in our example).

Program HPHI1U:

The program HPHI1U is a heuristic extension of DPHI1U that allows user-defined classes of objects to be the entities that are hierarchically clustered. The prompt to define such entities is exactly the same as given in HPCL1U. In addition to the 14 menu items from DPHI1U, HPHI1U includes 6 additional options:

```
MAIN MENU

  ⋮



15 ABSOLUTE SUM OF DEVIATIONS FROM MEDIAN WITH ORDER
   CONSTRAINTS PLUS LATER UNION CONSTRAINTS
16 SAME AS 15 BUT NO LATER UNION CONSTRAINTS
17 LARGEST ABSOLUTE DEVIATION FROM THE AVERAGE OF
   TWO EXTREMES PLUS LATER UNION CONSTRAINTS
18 SAME AS 17 BUT NO LATER UNION CONSTRAINTS
19 ULTRAMETRIC ORDER INCONSISTENCIES --- COMMON
   OBJECTS ARE NECESSARY
20 ULTRAMETRIC ORDER INCONSISTENCIES --- COMMON
   OBJECTS ARE UNNECESSARY
CHOICE?
```

In contrast (but analogous) to the least-squares options 13 and 14, options 15 and 16 use an absolute sum of deviations from the median (an L_1 norm), and 17 and 18 use the largest absolute deviation from the average of the two extreme proximities between a pair of subsets (an L_∞ norm). The two options 19 and 20 are the loss functions based on triple inconsistencies (OTIs) and quadruple inconsistencies (OQIs) to a base ultrametric, as discussed and illustrated in some detail in Section 3.2.1.

Program HPHI2U:

The program HPHI2U is an extension of HPHI1U to two-mode proximity matrices. The prompts to define the entities to be hierarchically clustered are exactly as in HPCL2U, and the data are entered as a rectangular matrix, also as in HPCL2U.

Object Sequencing/Seriation

The primary program for the sequencing of objects is DPSE1U with four others (DPSE2U, DPSE2R, HPSE1U, HPSE2U) interpretable as variants dealing with two-mode proximity data, the imposition of order restrictions on the rows and/or columns of such a two-mode matrix, or those allowing heuristic extensions for large object sets. We will illustrate explicitly the operation of DPSE1U with the prompts needed and the resulting output from carrying out three separate analyses. Using the symmetric proximity data on the digits in the file 'number.dat', the ten objects will be sequenced using both unidimensional coordinate estimation and the weighted gradient measure, and for the 'after' skew-symmetric matrix of Table 4.1, the sum of proximities will be maximized above the main diagonal of the reordered matrix. The latter skew-symmetric matrix will be in a file called 'tssa.dat' with format '(13F6.2)'; the contents of this input file are as follows:

```
  .00  -.58  -.86  -.90  -.34  -.40  -.96   .00 -1.00  -.88  -.94   .24  -.46
  .58   .00  -.02  -.40   .28   .38  -.90   .64  -.92  -.46  -.28   .46   .38
  .86   .02   .00  -.40   .44   .40  -.94   .74  -.96  -.36  -.24   .68   .40
  .90   .40   .40   .00   .68   .74  -.88   .78  -.92  -.34   .06   .84   .68
  .34  -.28  -.44  -.68   .00   .00  -.96   .36  -.98  -.78  -.62   .30   .08
  .40  -.38  -.40  -.74   .00   .00  -.96   .46  -.96  -.80  -.52   .40   .02
  .96   .90   .94   .88   .96   .96   .00   .98  -.28   .58   .88   .96   .96
  .00  -.64  -.74  -.78  -.36  -.46  -.98   .00  -.98  -.88  -.78  -.24  -.32
 1.00   .92   .96   .92   .98   .96   .28   .98   .00   .74   .90   .98   .96
  .88   .46   .36   .34   .78   .80  -.58   .88  -.74   .00   .28   .84   .78
  .94   .28   .24  -.06   .62   .52  -.88   .78  -.90  -.28   .00   .94   .52
 -.24  -.46  -.68  -.84  -.30  -.40  -.96   .24  -.98  -.84  -.94   .00  -.28
  .46  -.38  -.40  -.68  -.08  -.02  -.96   .32  -.96  -.78  -.52   .28   .00
```

The three analyses would in practice be carried out separately by three distinct runs of DPSE1U; however, we will indicate jointly what the responses to the given prompts would be in the listing below with the option for coordinate estimation given first, followed by the use of the weighted gradient measure, and finally, by the responses required for the analysis of the skew-symmetric proximity matrix.

```
MAIN MENU
    1 UNWEIGHTED GRADIENT WITHIN ROWS AND COLUMNS
        NUMBER OF STRICT CONSISTENCIES MINUS
        INCONSISTENCIES ABOVE THE MAIN DIAGONAL
    2 NUMBER OF CONSISTENCIES MINUS INCONSISTENCIES
        WITHIN ROWS LEFT AND RIGHT OF MAIN DIAGONAL
    3 DEFAYS CRITERION
        SUM OF SQUARES OF PROXIMITY SUMS TO MAIN
        DIAGONAL MINUS FROM MAIN DIAGONAL WITHIN ROWS
    4 PROXIMITY SUM
        ABOVE MAIN DIAGONAL
    5 WEIGHTED GRADIENT WITHIN ROWS AND COLUMNS
        WEIGHTED BY POSITIVE OR NEGATIVE DIFFERENCES
```

```
      6 GREENBERG PATTERN
        SAME AS 1 BUT REVERSED WITHIN THE COLUMN
        COMPARISONS
      7 EQUALLY SPACED COORDINATES-- PROXIMITY SUMS
        TO MAIN DIAGONAL MINUS FROM MAIN DIAGONAL
        WITHIN ROWS WEIGHTED BY INTEGER CONSTANTS
      8 GREENBERG PATTERN
        SAME AS 5 BUT REVERSED WITHIN THE COLUMN
        COMPARISONS
CHOICE?
[3] or
[5] or
[4]
```

{The options appropriate for a symmetric proximity matrix are 1 and 5, which provide the unweighted and weighted gradient measures; or 3 and 7, which give coordinate estimates, with option 7 restricted to those estimates that are equally spaced along a continuum. For historical reasons, option 3 is labeled as the Defays criterion after the individual who first suggested how such an optimization criterion could be used in the coordinate estimation context. For a skew-symmetric matrix, option 4 will maximize the above-diagonal sum (as will be seen in the relevant output file; this option also provides automatically the additional set of closed-form estimates for a set of coordinates); option 2 was not explicitly discussed in the text but maximizes the number of consistencies minus inconsistencies in the ordering of the proximities within the rows of a reordered skew-symmetric matrix to the left and right of the main diagonal. Finally, options 6 and 8 are the unweighted and weighted gradient measures for the alternative Greenberg pattern.}

```
1 FOR WITHIN ROW COMPARISONS ONLY
2 FOR WITHIN COLUMN COMPARISONS ONLY,
3 FOR   BOTH COMPARISONS
[3]
```

{This prompt is only given for the unweighted and weighted gradient measures in options 1 and 5; the comparisons could be restricted to be within the rows only, within the columns only, or both within the rows and the columns. The last option is chosen here in our illustration for the main menu option 5.}

```
NUMBER OF ROWS/COLUMNS IN THE INPUT MATRIX
[10] or
[10] or
[13]
```

```
0 FOR NO PRECEDENCE MATRIX TO BE READ IN AFTER THE DATA
1 IF A SQUARE PRECEDENCE MATRIX TO BE USED (2012 FORMAT;
```

0/1 WHERE A 1 IF THE ROW OBJECT MUST COME AFTER THE
COLUMN OBJECT
[0] or
[0] or
[0]

{If the object sequencing is to be restricted according to specified precedence constraints, an additional square matrix with 0/1 entries must appear after the proximity matrix in the input file (20I2 format for each row). If a 1 is placed at a particular row/column position in this matrix, then that particular row object would appear after the given column object in the object sequence generated. If a precedence matrix is used in a particular problem, it is also printed after the input matrix in the output file.}

{Given the size of the matrix that is to be input, the program at this point attempts to allocate the storage space needed. If successful, the phrase 'ARRAY ALLOCATION SUCCESSFUL' appears; if not, the phrase 'ARRAY ALLOCATION ERROR' is given, and the program terminates.}

FILE NAME FOR THE INPUT MATRIX
[number.dat] or
[number.dat] or
[tssa.dat]

FILE NAME FOR THE OUTPUT
[numser.out] or
[numserwg.out] or
[numsersk.out]

IS THE INPUT MATRIX SYMMETRIC AND READ AS A LOWER
TRIANGULAR MATRIX WITHOUT DIAGONAL ENTRIES OR AS
A COMPLETE SQUARE MATRIX
   1 LOWER TRIANGULAR
   2 COMPLETE SQUARE
   3 COMPLETE SQUARE AND SKEW-SYMMETRIC
   4 COMPLETE SQUARE AND SKEW-SYMMETRIC
      BUT USE SIGN INFORMATION ONLY
CHOICE?
[1] or
[1] or
[3]

{Symmetric proximity matrices can be given either in lower-triangular form or as a complete square matrix (options 1 and 2, respectively); a skew-symmetric matrix is entered as a square matrix (option 3), or if only the ±1 sign information from the skew-symmetric matrix is to be used, option 4 can be issued.}

APPENDIX: AVAILABLE PROGRAMS

PROVIDE FORMAT FOR THE INPUT MATRIX
[(9F4.3)] or
[(9F4.3)] or
[(13F6.2)]

{The input file must exist in the directory from which DPSE1U is run, and the given output file must not already exist in the directory. Otherwise, an error message appears at this point, and the program terminates.}

PROGRAM RUNNING

{No additional prompts are given until the processing is complete.}

The contents of the three output files are given below.

The output file when the coordinate estimation option 3 is used for the symmetric proximity matrix contained in 'number.dat' should be fairly self-explanatory. There is an indication of the optimization menu option selected, a listing of the input matrix, and the optimal sequence and cumulative objective function given in reverse order. For this example, the optimal sequencing would be: 8 10 9 7 6 4 5 3 2 1. The estimated coordinates are then given, along with the correlation between the proximities and reconstructed distances; the residual sum of squares is for the complete square matrix, and thus, each residual is counted twice because the original proximity matrix is symmetric. Finally, the proximity matrix is reprinted using the optimal object sequencing for the ordering of its rows and columns.

file 'numser.out':

OPTIMIZATION MENU OPTION   3

INPUT MATRIX

```
       0.421
       0.584    0.284
       0.709    0.346    0.354
       0.684    0.646    0.059    0.413
       0.804    0.588    0.671    0.429    0.409
       0.788    0.758    0.421    0.300    0.388    0.396
       0.909    0.630    0.796    0.592    0.742    0.400
       0.417
       0.821    0.791    0.367    0.804    0.246    0.671
       0.350    0.400
       0.850    0.625    0.808    0.263    0.683    0.592
       0.296    0.459    0.392
```

```
CUMULATIVE OBJECTIVE FUNCTION      PERMUTATION IN REVERSE ORDER
           130.797                              1
            87.632                              2
            69.595                              3
            62.794                              5
            60.568                              4
            60.247                              6
            59.538                              7
            55.586                              9
            44.972                             10
            28.569                              8

COORDINATES BASED ON DEFAYS CRITERION
    -0.535     -0.405     -0.326     -0.199     -0.084      0.057      0.149
     0.261      0.425      0.657


CORRELATION BETWEEN PROXIMITIES AND RECONSTRUCTED DISTANCES IS   0.736

RESIDUAL SUM OF SQUARES FOR SQUARE MATRIX        3.9197

REORDERED MATRIX
           0.000      0.459      0.400      0.417      0.400      0.592
           0.742      0.796      0.630      0.909

           0.459      0.000      0.392      0.296      0.592      0.263
           0.683      0.808      0.625      0.850

           0.400      0.392      0.000      0.350      0.671      0.804
           0.246      0.367      0.791      0.821

           0.417      0.296      0.350      0.000      0.396      0.300
           0.388      0.421      0.758      0.788

           0.400      0.592      0.671      0.396      0.000      0.429
           0.409      0.671      0.588      0.804

           0.592      0.263      0.804      0.300      0.429      0.000
           0.413      0.354      0.346      0.709

           0.742      0.683      0.246      0.388      0.409      0.413
           0.000      0.059      0.646      0.684

           0.796      0.808      0.367      0.421      0.671      0.354
           0.059      0.000      0.284      0.584

           0.630      0.625      0.791      0.758      0.588      0.346
           0.646      0.284      0.000      0.421

           0.909      0.850      0.821      0.788      0.804      0.709
           0.684      0.584      0.421      0.000
```

In comparison with the previous output file, the use of the weighted gradient option 5 includes an indication that both within-row and within-column comparisons are being used; the optimal index for the weighted gradient measure is provided along with the component terms that define it.

file 'numserwg.out':

```
OPTIMIZATION MENU OPTION     5

BOTH WITHIN ROW AND COLUMN COMPARISONS USED

INPUT MATRIX

        0.421

        0.584       0.284

        0.709       0.346       0.354

        0.684       0.646       0.059       0.413

        0.804       0.588       0.671       0.429       0.409

        0.788       0.758       0.421       0.300       0.388       0.396

        0.909       0.630       0.796       0.592       0.742       0.400
        0.417

        0.821       0.791       0.367       0.804       0.246       0.671
        0.350       0.400

        0.850       0.625       0.808       0.263       0.683       0.592
        0.296       0.459       0.392


    CUMULATIVE OBJECTIVE FUNCTION     PERMUTATION IN REVERSE ORDER
                41.822                           1
                41.822                           2
                37.560                           3
                30.690                           5
                23.595                           4
                16.067                           6
                12.802                           7
                 5.153                           9
                 1.591                          10
                 0.000                           8

    INDEX OF GRADIENT COMPARISON IS    41.822

    CONSISTENCIES     49.109 INCONSISTENCIES      7.287 DIFFERENCE 41.822 SUM 56.396
    RATIO 0.742

    REORDERED MATRIX

    {The reordered matrix has not been printed.}
```

For the analysis of the skew-symmetric matrix in the file 'tssa.dat', most of the output file parallels the file given above for the two other analyses. The optimal index when maximizing the above-diagonal sum of proximities is given in the form of the 'average above-diagonal proximity' for the reordered matrix. In addition to these results and because a skew-symmetric matrix is being analyzed, the closed-form coordinate estimates are given along with the object ordering they would induce; for a restriction to equally spaced coordinates, the multiplicative constant (alpha) is provided. The correlations are listed between the skew-symmetric proximities and the reconstructed distances, along with the residual sum of squares for both the cases of equally spaced and estimated coordinates; another reordered proximity matrix is given using the latter ordering of the estimated coordinates.

OBJECT SEQUENCING/SERIATION

file 'numsersk.out':

```
OPTIMIZATION MENU OPTION    4

INPUT MATRIX
          0.000   -0.580   -0.860   -0.900   -0.340   -0.400
         -0.960    0.000   -1.000   -0.880   -0.940    0.240
         -0.460
          0.580    0.000   -0.020   -0.400    0.280    0.380
         -0.900    0.640   -0.920   -0.460   -0.280    0.460
          0.380
          0.860    0.020    0.000   -0.400    0.440    0.400
         -0.940    0.740   -0.960   -0.360   -0.240    0.680
          0.400
          0.900    0.400    0.400    0.000    0.680    0.740
         -0.880    0.780   -0.920   -0.340    0.060    0.840
          0.680
          0.340   -0.280   -0.440   -0.680    0.000    0.000
         -0.960    0.360   -0.980   -0.780   -0.620    0.300
          0.080
          0.400   -0.380   -0.400   -0.740    0.000    0.000
         -0.960    0.460   -0.960   -0.800   -0.520    0.400
          0.020
          0.960    0.900    0.940    0.880    0.960    0.960
          0.000    0.980   -0.280    0.580    0.880    0.960
          0.960
          0.000   -0.640   -0.740   -0.780   -0.360   -0.460
         -0.980    0.000   -0.980   -0.880   -0.780   -0.240
         -0.320
          1.000    0.920    0.960    0.920    0.980    0.960
          0.280    0.980    0.000    0.740    0.900    0.980
          0.960
          0.880    0.460    0.360    0.340    0.780    0.800
         -0.580    0.880   -0.740    0.000    0.280    0.840
          0.780
          0.940    0.280    0.240   -0.060    0.620    0.520
         -0.880    0.780   -0.900   -0.280    0.000    0.940
          0.520
         -0.240   -0.460   -0.680   -0.840   -0.300   -0.400
         -0.960    0.240   -0.980   -0.840   -0.940    0.000
         -0.280
          0.460   -0.380   -0.400   -0.680   -0.080   -0.020
         -0.960    0.320   -0.960   -0.780   -0.520    0.280
          0.000


     CUMULATIVE OBJECTIVE FUNCTION    PERMUTATION IN REVERSE ORDER
                47.420                             8
                40.260                            12
                33.340                             1
                26.020                            13
                21.240                             5
                16.500                             6
                11.740                             2
                 8.760                             3
                 5.860                            11
                 3.740                             4
                 1.600                            10
                 0.280                             7
                 0.000                             9

AVERAGE ABOVE DIAGONAL PROXIMITY IS       0.608
AVERAGE BELOW DIAGONAL PROXIMITY IS      -0.608
```

REORDERED MATRIX

0.000	0.280	0.740	0.920	0.900	0.960
0.920	0.960	0.980	0.960	1.000	0.980
0.980					
-0.280	0.000	0.580	0.880	0.880	0.940
0.900	0.960	0.960	0.960	0.960	0.960
0.980					
-0.740	-0.580	0.000	0.340	0.280	0.360
0.460	0.800	0.780	0.780	0.880	0.840
0.880					
-0.920	-0.880	-0.340	0.000	0.060	0.400
0.400	0.740	0.680	0.680	0.900	0.840
0.780					
-0.900	-0.880	-0.280	-0.060	0.000	0.240
0.280	0.520	0.620	0.520	0.940	0.940
0.780					
-0.960	-0.940	-0.360	-0.400	-0.240	0.000
0.020	0.400	0.440	0.400	0.860	0.680
0.740					
-0.920	-0.900	-0.460	-0.400	-0.280	-0.020
0.000	0.380	0.280	0.380	0.580	0.460
0.640					
-0.960	-0.960	-0.800	-0.740	-0.520	-0.400
-0.380	0.000	0.000	0.020	0.400	0.400
0.460					
-0.980	-0.960	-0.780	-0.680	-0.620	-0.440
-0.280	0.000	0.000	0.080	0.340	0.300
0.360					
-0.960	-0.960	-0.780	-0.680	-0.520	-0.400
-0.380	-0.020	-0.080	0.000	0.460	0.280
0.320					
-1.000	-0.960	-0.880	-0.900	-0.940	-0.860
-0.580	-0.400	-0.340	-0.460	0.000	0.240
0.000					
-0.980	-0.960	-0.840	-0.840	-0.940	-0.680
-0.460	-0.400	-0.300	-0.280	-0.240	0.000
0.240					
-0.980	-0.980	-0.880	-0.780	-0.780	-0.740
-0.640	-0.460	-0.360	-0.320	0.000	-0.240
0.000					

CLOSED FORM RESULTS FOR SKEW-SYMMETRIC MATRICES

OBJECT REORDERING
  9 7 10 4 11 3 2 6 5 13 12 1 8

COORDINATES
 -0.814 -0.745 -0.391 -0.257 -0.209 -0.049 0.020 0.268 0.282 0.286
  0.514   0.545   0.551

MULTIPLICATIVE CONSTANT ALPHA   0.116

EQUALLY SPACED
 CORRELATION BETWEEN PROXIMITIES AND RECONSTRUCTED DISTANCES IS 0.788
 RESIDUAL SUM OF SQUARES FOR SQUARE MATRIX   7.7050

ESTIMATED COORDINATES
 CORRELATION BETWEEN PROXIMITIES AND RECONSTRUCTED DISTANCES IS 0.896
 RESIDUAL SUM OF SQUARES FOR SQUARE MATRIX   4.7380

REORDERED MATRIX

{reordered matrix not printed}

Program HPSE1U:

The program HPSE1U is a heuristic extension of DPSE1U with all the same options but allows user-defined classes of objects to be the entities that are sequenced. The prompt to define such entities is exactly the same as that given by HPCL1U (and HPHI1U).

Program DPSE2U:

The program DPSE2U is a variant of DPSE1U to deal with two-mode proximity data, which are entered as a rectangular matrix (as in HPHI2U or HPCL2U). The main menu options are fewer than in DPSE1U because coordinate estimation options or skew-symmetric matrices are not included. We give the beginning prompts below in running DPSE2U.

```
MAIN MENU
    1 UNWEIGHTED GRADIENT WITHIN ROWS AND COLUMNS
        NUMBER OF STRICT CONSISTENCIES MINUS
        INCONSISTENCIES ABOUT THE MAIN DIAGONAL
    2 WEIGHTED GRADIENT WITHIN ROWS AND COLUMNS
        WEIGHTED BY POSITIVE OR NEGATIVE DIFFERENCES
    3 UNWEIGHTED GRADIENT WITHIN ROWS AND COLUMNS
        NUMBER OF INCONSISTENCIES TO BE MINIMIZED
    4 WEIGHTED GRADIENT WITHIN ROWS AND COLUMNS
        WEIGHTED INCONSISTENCIES TO BE MINIMIZED
CHOICE?
[?]
```

{The optimization options 1 and 2 are the unweighted and weighted gradient measures and are analogous to those available in DPSE1U. Options 3 and 4 allow the minimization of just the inconsistencies in the unweighted and weighted gradient measures.}

```
1 FOR WITHIN ROW COMPARISONS ONLY
2 FOR WITHIN COLUMN COMPARISONS ONLY
3 FOR BOTH COMPARISONS -- THESE CHOICES ARE
   FOR THE SQUARE DERIVED MATRIX
[?]
```

{This option allows comparisons to be restricted to the rows or to the columns of the derived square proximity matrix; the given rectangular proximity matrix is embedded in the upper-right and lower-left portions of the derived square proximity matrix with missing entries between the row objects and between the column objects.}

```
1 FOR WITHIN ROW COMPARISONS ONLY
2 FOR WITHIN COLUMN COMPARISONS ONLY
3 FOR BOTH COMPARISONS -- THESE CHOICES ARE
  FOR THE ORIGINAL RECTANGULAR MATRIX
[?]
```

{This option allows comparisons to be restricted to the rows or to the columns of the original rectangular proximity matrix.}

```
NUMBER OF ROWS IN THE INPUT MATRIX
[?]

NUMBER OF COLUMNS IN THE INPUT MATRIX
[?]

0 FOR NO PRECEDENCE MATRIX TO BE READ IN AFTER THE DATA
1 IF A SQUARE PRECEDENCE MATRIX TO BE USED (2012 FORMAT;
  0/1 WHERE A 1 IF THE ROW OBJECT MUST COME AFTER THE
  COLUMN OBJECT
[?]
```

{If selected, the precedence matrix that comes after the data matrix in the input file is square and of size equal to the sum of the number of row and column objects (thus, the column objects from the original rectangular proximity matrix are labeled starting at 1 + number of row objects, and ending at number of column objects + number of row objects). A 1 is placed in this square matrix to indicate that the specific object that this row refers to, which may be a row or column object in the original rectangular proximity matrix, must come after the specific object that the column refers to, which may also be a row or column object in the original rectangular proximity matrix.}

Program HPSE2U:

The program HPSE2U is a heuristic extension of DPSE2U with the same options that allow user-specified classes of objects to be the entities that are sequenced. The prompts to define such entities are exactly the same as those given by HPCL2U (or HPHI2U).

Program DPSE2R:

The program DPSE2R is an extension of DPSE2U that allows order constraints to be imposed on the column objects only, or on both the row objects and the column objects. In addition, if order constraints are imposed on both the row and the column objects, the one additional optimization option under the main menu is allowed that minimizes the DHG sequence comparison measure, i.e.,

```
MAIN MENU

    ⋮

5 DHG SEQUENCE COMPARISON MEASURE TO BE
  MINIMIZED -- ROW AND COLUMN ORDER CONSTRAINTS
[?]
```

The constraints are imposed by response to the prompts

```
ORDER RESTRICTIONS?
1 COLUMN ORDER RESTRICTIONS ONLY
2 ROW AND COLUMN ORDER RESTRICTIONS
[?]

COLUMN ORDER RESTRICTION
1 IDENTITY ORDER
2 TO BE READ IN
[?]
```

If '2' is the option used in the 'order restrictions?' prompt, we get:

```
ROW ORDER RESTRICTION
1 IDENTITY ORDER
2 TO BE READ IN
[?]
```

If the column order or the row order is to be input, prompts will appear of the form

```
INPUT COLUMN ORDER IN INTEGER FORM SEPARATED BY BLANKS
[...integer labels for the column order...]

INPUT ROW ORDER IN INTEGER FORM SEPARATED BY BLANKS
[...integer label for the row order...]
```

Object Sequencing by Constructing Optimal Paths

The program DPSEPH constructs an optimal path (either linear or circular) among the objects in a given set based on a proximity matrix. We illustrate the operation of DPSEPH with the prompts needed to find the minimum length (linear) path for the symmetric proximity data on the digits in the file 'number.dat', and include below the relevant output matrix 'numph.out'.

```
MAIN MENU
  1 MINIMUM PATH
  2 MINIMAX PATH
  3 MAXIMUM PATH
  4 MAXIMIN PATH
  5 MINIMUM CIRCULAR PATH
  6 MINIMAX CIRCULAR PATH
  7 MAXIMUM CIRCULAR PATH
  8 MAXIMIN CIRCULAR PATH
CHOICE?
[1]
```

{In constructing either an optimal linear or circular path, the optimization criterion for the path 'length' may be one of four types: the sum of proximities on the edges can be either minimized or maximized, the maximum proximity on an edge can be minimized, or the minimum proximity on an edge can be maximized. In our example, we specify the sum of edges in the path to be at a minimum, so option 1 is chosen.}

```
NUMBER OF ROWS/COLUMNS IN THE INPUT MATRIX
[10]

0 FOR NO PRECEDENCE MATRIX TO BE READ IN AFTER THE DATA
1 IF A SQUARE PRECEDENCE MATRIX TO BE USED (2012 FORMAT;
  0/1 WHERE A 1 IF THE ROW OBJECT MUST COME AFTER THE
  COLUMN OBJECT
[0]
```

{At this point, the array allocation process is attempted with the usual messages given if successful or unsuccessful.}

```
FILE NAME FOR THE INPUT MATRIX
[number.dat]

FILE NAME FOR THE OUTPUT
[numph.out]

IS THE INPUT MATRIX SYMMETRIC AND READ AS A LOWER
TRIANGULAR MATRIX WITHOUT DIAGONAL ENTRIES OR AS
A COMPLETE SQUARE MATRIX
  1 LOWER TRIANGULAR
  2 COMPLETE SQUARE
[1]
```

{As noted in the text, if the proximity matrix entered is square and not symmetric, optimal directed linear or circular paths would be constructed.}

CONSTRUCTING OPTIMAL PATHS

PROVIDE FORMAT FOR THE INPUT MATRIX
[(9F4.3)]

PROGRAM RUNNING

The contents of the output file 'numph.out' are given below, where the items should be self-explanatory. The optimal linear path is in the order 8 6 4 10 7 9 5 3 2 1, with a sum of proximities on the edges equal to 2.748.

```
OPTIMIZATION MENU OPTION    1
INPUT MATRIX
            0.421

            0.584       0.284

            0.709       0.346       0.354

            0.684       0.646       0.059       0.413

            0.804       0.588       0.671       0.429       0.409

            0.788       0.758       0.421       0.300       0.388       0.396

            0.909       0.630       0.796       0.592       0.742       0.400
            0.417

            0.821       0.791       0.367       0.804       0.246       0.671
            0.350       0.400

            0.850       0.625       0.808       0.263       0.683       0.592
            0.296       0.459       0.392


CUMULATIVE OBJECTIVE     2.748    LAST OBJECT        1
CUMULATIVE OBJECTIVE     2.327    LAST OBJECT        2
CUMULATIVE OBJECTIVE     2.043    LAST OBJECT        3
CUMULATIVE OBJECTIVE     1.984    LAST OBJECT        5
CUMULATIVE OBJECTIVE     1.738    LAST OBJECT        9
CUMULATIVE OBJECTIVE     1.388    LAST OBJECT        7
CUMULATIVE OBJECTIVE     1.092    LAST OBJECT       10
CUMULATIVE OBJECTIVE     0.829    LAST OBJECT        4
CUMULATIVE OBJECTIVE     0.400    LAST OBJECT        6
BEGINNING OBJECTIVE      0.000    FIRST OBJECT       8

REORDERED MATRIX
            0.000       0.400       0.592       0.459       0.417       0.400
            0.742       0.796       0.630       0.909

            0.400       0.000       0.429       0.592       0.396       0.671
            0.409       0.671       0.588       0.804

            0.592       0.429       0.000       0.263       0.300       0.804
            0.413       0.354       0.346       0.709

            0.459       0.592       0.263       0.000       0.296       0.392
            0.683       0.808       0.625       0.850

            0.417       0.396       0.300       0.296       0.000       0.350
            0.388       0.421       0.758       0.788

            0.400       0.671       0.804       0.392       0.350       0.000
```

```
        0.246      0.367       0.791       0.821

        0.742      0.409       0.413       0.683       0.388       0.246
        0.000      0.059       0.646       0.684

        0.796      0.671       0.354       0.808       0.421       0.367
        0.059      0.000       0.284       0.584

        0.630      0.588       0.346       0.625       0.758       0.791
        0.646      0.284       0.000       0.421

        0.909      0.804       0.709       0.850       0.788       0.821
        0.684      0.584       0.421       0.000
```

Constructing Ordered Partitions

The main program for constructing optimal ordered partitions is called DPOP1U. A heuristic extension is available in HPOP1U with the same options as in DPOP1U, but which allows user-defined classes of objects to be the entities that are partitioned and the resulting classes of these entities to be sequenced. The prompt to define these entities is the same as given in HPCL2U, HPHI2U, or HPSE2U. We will illustrate the use of DPOP1U and the various prompts given, again with the digit data in the file 'number.dat', and will construct optimal ordered partitions into 2 through 10 classes based on coordinate estimation. The output file 'numop.out' will be included below.

```
MAIN MENU
  1 UNWEIGHTED GRADIENT WITHIN ROWS AND COLUMNS
  2 WEIGHTED GRADIENT WITHIN ROWS AND COLUMNS
  3 DEFAYS CRITERION
  4 PROXIMITY SUM
  5 SKEW-SYMMETRIC COORDINATES
CHOICE?
[3]
```

{As discussed in the text, options 1, 2, and 3 are for symmetric proximity matrices; options 4 and 5 are for skew-symmetric matrices.}

```
MAXIMUM NUMBER OF SUBSETS TO BE USED
[10]

NUMBER OF ROWS/COLUMNS
[10]
```

{If this latter response is less than 3 or greater than 30, an error message appears that the 'PROGRAM SIZE IS OUT OF RANGE', and the program terminates. Otherwise, array allocation is attempted and the usual success or error messages are given.}

CONSTRUCTING ORDERED PARTITIONS

```
FILE NAME FOR THE INPUT MATRIX
[number.dat]

FILE NAME FOR THE OUTPUT MATRIX
[numop.out]

IS THE INPUT MATRIX SYMMETRIC AND READ A LOWER
TRIANGULAR MATRIX WITHOUT DIAGONAL ENTRIES OR AS
A COMPLETE SQUARE MATRIX
  1 LOWER TRIANGULAR
  2 COMPLETE SQUARE
  3 COMPLETE SQUARE AND SKEW SYMMETRIC
  4 COMPLETE SQUARE AND SKEW-SYMMETRIC
     BUT USE SIGN INFORMATION ONLY
CHOICE?
[1]

PROVIDE FORMAT FOR THE INPUT MATRIX
[(9F4.3)]

PROGRAM RUNNING
```

{Again, no response to this last prompt is necessary.}

The contents of the output file 'numop.out' are given below, where the various items should be self-explanatory for each of the optimal ordered partitions listed.

```
   OPTIMIZATION MENU OPTION    3
INPUT MATRIX
         0.421
         0.584    0.284
         0.709    0.346    0.354
         0.684    0.646    0.059    0.413
         0.804    0.588    0.671    0.429    0.409
         0.788    0.758    0.421    0.300    0.388    0.396
         0.909    0.630    0.796    0.592    0.742    0.400
         0.417
         0.821    0.791    0.367    0.804    0.246    0.671
         0.350    0.400
         0.850    0.625    0.808    0.263    0.683    0.592
         0.296    0.459    0.392
```

NUMBER OF SUBSETS   1

SUBSET ADDED 10 9 8 7 6 5 4 3 2 1
INDEX FOR SUBSET   0.0000   CUMULATIVE INDEX   0.0000

SUMMARY SUBSET MEMBERSHIP FOR THE OBJECTS
 1 1 1 1 1 1 1 1 1 1

COORDINATES FOR THE SUBSETS
  0.000

RESIDUAL SUM OF SQUARES   30.079

NUMBER OF SUBSETS   2

SUBSET ADDED 5 4 3 2 1
INDEX FOR SUBSET   47.9447   CUMULATIVE INDEX   95.8893

SUBSET ADDED 10 9 8 7 6
INDEX FOR SUBSET   47.9447   CUMULATIVE INDEX   47.9447

SUMMARY SUBSET MEMBERSHIP FOR THE OBJECTS
 2 2 2 2 2 1 1 1 1 1

COORDINATES FOR THE SUBSETS
 -0.310   0.310

RESIDUAL SUM OF SQUARES   10.901

NUMBER OF SUBSETS   3

SUBSET ADDED 10 9 8 7
INDEX FOR SUBSET   53.5897   CUMULATIVE INDEX   115.7492

SUBSET ADDED 6 5 4 3
INDEX FOR SUBSET   3.6557   CUMULATIVE INDEX   62.1595

SUBSET ADDED 2 1
INDEX FOR SUBSET   58.5038   CUMULATIVE INDEX   58.5038

SUMMARY SUBSET MEMBERSHIP FOR THE OBJECTS
 1 1 2 2 2 2 3 3 3 3

COORDINATES FOR THE SUBSETS
 -0.541 -0.096   0.366

RESIDUAL SUM OF SQUARES   6.929

NUMBER OF SUBSETS   4

SUBSET ADDED 2 1
INDEX FOR SUBSET   58.5038   CUMULATIVE INDEX   123.1315

SUBSET ADDED 5 4 3
INDEX FOR SUBSET   7.2572   CUMULATIVE INDEX   64.6278

SUBSET ADDED 7 6
INDEX FOR SUBSET   4.0044   CUMULATIVE INDEX   57.3706

SUBSET ADDED 10 9 8
INDEX FOR SUBSET   53.3661   CUMULATIVE INDEX   53.3661

CONSTRUCTING ORDERED PARTITIONS

SUMMARY SUBSET MEMBERSHIP FOR THE OBJECTS
  4 4 3 3 3 2 2 1 1 1

COORDINATES FOR THE SUBSETS
  -0.422 -0.141   0.156   0.541

RESIDUAL SUM OF SQUARES   5.453

NUMBER OF SUBSETS   5

SUBSET ADDED 10 9 8
INDEX FOR SUBSET   53.3661   CUMULATIVE INDEX   126.1487

SUBSET ADDED 7 6
INDEX FOR SUBSET   4.0044   CUMULATIVE INDEX   72.7825

SUBSET ADDED 5 4
INDEX FOR SUBSET   2.1177   CUMULATIVE INDEX   68.7781

SUBSET ADDED 3 2
INDEX FOR SUBSET   23.4955   CUMULATIVE INDEX   66.6604

SUBSET ADDED 1
INDEX FOR SUBSET   43.1649   CUMULATIVE INDEX   43.1649

SUMMARY SUBSET MEMBERSHIP FOR THE OBJECTS
  1 2 2 3 3 4 4 5 5 5

COORDINATES FOR THE SUBSETS
  -0.657 -0.343 -0.103   0.141   0.422

RESIDUAL SUM OF SQUARES   4.849

{The results for 6 to 9 subsets have not been printed.}

NUMBER OF SUBSETS 10

SUBSET ADDED 8
INDEX FOR SUBSET   28.5690   CUMULATIVE INDEX   130.7972

SUBSET ADDED 10
INDEX FOR SUBSET   16.4025   CUMULATIVE INDEX   102.2282

SUBSET ADDED 9
INDEX FOR SUBSET   10.6146   CUMULATIVE INDEX   85.8257

SUBSET ADDED 7
INDEX FOR SUBSET   3.9521   CUMULATIVE INDEX   75.2111

SUBSET ADDED 6
INDEX FOR SUBSET   0.7090   CUMULATIVE INDEX   71.2590

SUBSET ADDED 4
INDEX FOR SUBSET   0.3204   CUMULATIVE INDEX   70.5500

SUBSET ADDED 5
INDEX FOR SUBSET   2.2261   CUMULATIVE INDEX   70.2296

SUBSET ADDED 3
INDEX FOR SUBSET   6.8017   CUMULATIVE INDEX   68.0036

SUBSET ADDED 2
INDEX FOR SUBSET   18.0370   CUMULATIVE INDEX   61.2019

```
SUBSET ADDED   1
INDEX FOR SUBSET       43.1649   CUMULATIVE INDEX       43.1649

SUMMARY SUBSET MEMBERSHIP FOR THE OBJECTS
  1  2  3  5  4  6  7 10  8  9

COORDINATES FOR THE SUBSETS
  -0.657  -0.425  -0.261  -0.149  -0.057   0.084   0.199   0.326   0.405   0.535

RESIDUAL SUM OF SQUARES         3.920
```

Endnote

[45] The programs discussed in this appendix are all written in what might be referred to as a straightforward programming style and with a very transparent use of nested 'DO' loops. It is possible, however, to increase the speed of computation dramatically by more elegant uses of the architecture of the combinatorial tasks under consideration and by considering how these relate to some of the elemental operations available under Fortran90. Work along these lines has been done by Bart Jan van Os from Leiden University and will be reported in the literature in the near future.

Bibliography

Adelson, R. M., Norman, J. M., and Laporte, G. (1976). A dynamic programming formulation with diverse applications. *Operational Research Quarterly*, *27*, 119–121.

Alpert, C. J., and Kahng, A. B. (1995). Multiway partitioning via geometric embeddings, orderings, and dynamic programming. *IEEE Transactions on Computer-Aided Design of Integrated Circuits and Systems*, *14*, 1342–1357.

Alpert, C. J., and Kahng, A. B. (1997). Splitting an ordering into a partition to minimize diameter. *Journal of Classification*, *14*, 51–74.

Arabie, P., and Hubert, L. J. (1990). The bond energy algorithm revisited. *IEEE Transactions on Systems, Man, and Cybernetics*, *20*, 268–274.

Arabie, P., and Hubert, L. J. (1996). An overview of combinatorial data analysis. In P. Arabie, L. Hubert, and G. De Soete, Eds., *Clustering and Classification*. Singapore: World Scientific, 5–63.

Barthélemy, J.-P., and Guénoche, A. (1991). *Trees and Proximity Representations*. Chichester, U.K.: Wiley.

Barthélemy, J.-P., Hudry, O., Isaak, G., Roberts, F. S., and Tesman, B. (1995). The reversing number of a digraph. *Discrete Applied Mathematics*, *60*, 39–76.

Batagelj, V., Korenjak-Černe, S., and Klavžar, S. (1994). Dynamic programming and convex clustering. *Algorithmica*, *11*, 93–103.

Bellman, R. (1962). Dynamic programming treatment of the traveling salesman problem. *Journal of the Association for Computing Machinery*, *9*, 61–63.

Bezembinder, T., and van Acker, P. (1980). Intransitivity in individual and social choice. In E. D. Lantermann and H. Feger, Eds., *Similarity and Choice*, Bern: Huber, 208–233.

Bowman, V. J., and Colantoni, C. S. (1973). Majority rule under transitivity constraints. *Management Science*, *19*, 1029–1041.

Bowman, V. J., and Colantoni, C. S. (1974). Further comments on majority rule under transitivity constraints. *Management Science*, *20*, 1441.

Carroll, J. D., and Pruzansky, S. (1980). Discrete and hybrid scaling models. In E. D. Lantermann and H. Feger, Eds., *Similarity and Choice*. Bern: Huber, 108–139.

Chandon, J.-L., and De Soete, G. (1984). Fitting a least squares ultrametric to dissimilarity data: Approximation versus optimization. In E. Diday, M. Jambu, L. Lebart, J. Pages and R. Tomassone, eds., *Data Analysis and Informatics, III*. Amsterdam: North-Holland, 213–221.

Chandon, J.-L., Lemaire, J., and Pouget, J. (1980). Construction de l' ultramétrique la plus proche d' une dissimilarité au sens des moindres carrés. *RAIRO Recherche Opérationelle, 14*, 157–170.

Charon, I., Guénoche, A., Hudry, O., and Woirgard, F. (1997). New results on the computation of median orders. *Discrete Mathematics, 165/166*, 139–153.

Charon, I., Hudry O., and Woirgard, F. (1996). Ordres médians et ordres de Slater des tournois. *Mathémathiques, Informatique et Sciences Humaines, 34*, 23–56.

Chepoi, V., and Fichet, B. (in press). l_∞-approximation via subdominants. *Mathematical Social Sciences*.

Coombs, C. H. (1964). *A Theory of Data*. New York: Wiley.

Critchley, F. (1994). On exchangeability-based equivalence relations induced by strongly Robinson and, in particular, by quadripolar Robinson dissimilarity matrices. In B. van Cutsem, Ed., *Classification and Dissimilarity Analysis*. New York: Springer-Verlag, 5–65.

Daws, J. T. (1996). The analysis of free-sorting data: Beyond pairwise occurrences. *Journal of Classification, 13*, 57–80.

Day, W. H. E. (1996). Complexity theory: An introduction for practitioners of classification. In P. Arabie, L. Hubert, and G. De Soete, Eds., *Clustering and Classification*. Singapore: World Scientific, 199–233.

Defays, D. (1978). A short note on a method of seriation. *British Journal of Mathematical and Statistical Psychology, 31*, 49–53.

Delcoigne, A., and Hansen, P. (1975). Sequence comparison by dynamic programming. *Biometrika, 62*, 661–664.

De Soete, G. (1984). A least squares algorithm for fitting an ultrametric tree to a dissimilarity matrix. *Pattern Recognition Letters, 2*, 133–137.

De Soete, G., DeSarbo, W. S., Furnas, G. W., and Carroll, J. D. (1984a). Tree representation of rectangular proximity matrices. In E. Degreef and J. van Buggenhaut, Eds., *Trends in Mathematical Psychology*. Amsterdam: North-Holland, 377–392.

De Soete, G., DeSarbo, W. S., Furnas, G. W., and Carroll, J. D. (1984b). The estimation of ultrametric and path length trees from rectangular proximity data. *Psychometrika*, *49*, 289–310.

Diday, E. (1986). Orders and overlapping clusters by pyramids. In J. de Leeuw, W. J. Heiser, J. Meulman, and F. Critchley, Eds., *Multidimensional Data Analysis*. Leiden: DSWO Press, 201–234.

Durand, C., and Fichet, B. (1988). One-to-one correspondences in pyramidal representations: A unified approach. In H. H. Bock, Ed., *Classification and Related Methods of Data Analysis*. New York: Springer-Verlag, 85–90.

Farach, M., Kannan, S., and Warnow, T. (1995). A robust model for finding optimal evolutionary trees. *Algorithmica*, *13*, 155–179.

Fisher, W. D. (1958). On grouping for maximum heterogeneity. *Journal of the American Statistical Association*, *53*, 789–798.

Flueck, J. A., and Korsh, J. F. (1974). A branch search algorithm for maximum likelihood paired comparison ranking. *Biometrika*, *61*, 621–626.

Gelfand, A. E. (1971). Rapid seriation methods with archaeological applications. In F. R. Hodson, D. G. Kendall, and P. Tăutu, Eds., *Mathematics in the Archaeological and Historical Sciences*. Edinburgh: Edinburgh University Press, 186–201.

Gordon, A. D. (1973). A sequence-comparison statistic and algorithm. *Biometrika*, *60*, 197–200.

Greenberg, M. G. (1965). A method of successive cumulations for scaling of pair-comparison judgments. *Psychometrika*, *30*, 441–448.

Groenen, P. J. F. (1993). *The Majorization Approach to Multidimensional Scaling: Some Problems and Extensions*. Leiden, The Netherlands: DSWO Press.

Guttman. L. (1968). A general nonmetric technique for finding the smallest coordinate space for a configuration of points. *Psychometrika*, *33*, 469–506.

Hall, K. M. (1970). An r-dimensional quadratic placement algorithm. *Management Science*, *17*, 219–229.

Hansen, P., and Jaumard, B. (1997). Cluster analysis and mathematical programming. *Mathematical Programming*, *79*, 191–215.

Hartigan, J. A. (1967). Representation of similarity matrices by trees. *Journal of the American Statistical Association*, *62*, 1140–1158.

Hartigan, J. A. (1975). *Clustering Algorithms*. New York: Wiley.

Heiser, W. J. (1981). *Unfolding Analysis of Proximity Data*. Unpublished doctoral dissertation. Leiden, The Netherlands: Leiden University.

Held, M., and Karp, R. M. (1962). A dynamic programming approach to sequencing problems. *Journal of the Society for Industrial and Applied Mathematics, 10*, 196–210.

Hillier, F. S., and Lieberman, G. J. (1990). *Introduction to Mathematical Programming.* New York: McGraw-Hill.

Hodson, F. R., Kendall, D. G., and Tăutu, P., Eds. (1971). *Mathematics in the Archaeological and Historical Sciences.* Edinburgh: Edinburgh University Press.

Hubert, L. J. (1974a). Some applications of graph theory to clustering. *Psychometrika, 39*, 283–309.

Hubert, L. J. (1974b). Spanning trees and aspects of clustering. *British Journal of Mathematical and Statistical Psychology, 27*, 14–28.

Hubert, L. J. (1976). Seriation using asymmetric proximity measures. *British Journal of Mathematical and Statistical Psychology, 29*, 32–52.

Hubert, L. J. (1987). *Assignment Methods in Combinatorial Data Analysis.* New York: Marcel Dekker.

Hubert, L. J., and Arabie, P. (1986). Unidimensional scaling and combinatorial optimization. In J. de Leeuw, W. Heiser, J. Meulman, and F. Critchley, Eds., *Multidimensional Data Analysis.* Leiden, the Netherlands: DSWO Press, 181–196.

Hubert, L. J., and Arabie, P. (1994). The analysis of proximity matrices through sums of matrices having (anti-)Robinson forms. *British Journal of Mathematical and Statistical Psychology, 47*, 1–40.

Hubert, L. J., and Arabie, P. (1995a). The approximation of two-mode proximity matrices by sums of order-constrained matrices. *Psychometrika, 60*, 573–605.

Hubert, L. J., and Arabie, P. (1995b). Iterative projection strategies for the least-squares fitting of tree structures to proximity data. *British Journal of Mathematical and Statistical Psychology, 48*, 281–317.

Hubert, L. J., Arabie, P., and Meulman, J. (1997a). Linear and circular unidimensional scaling for symmetric proximity matrices. *British Journal of Mathematical and Statistical Psychology, 50*, 253–284.

Hubert, L. J., Arabie, P., and Meulman J. (1997b). The construction of globally optimal ordered partitions. In B. Mirkin, F. R. McMorris, F. S. Roberts, and A. Rzhetsky, Eds., *Mathematical Hierarchies and Biology* (DIMACS Series in Discrete Mathematics and Theoretical Computer Science). Providence, RI: American Mathematical Society, 299–312.

Hubert, L. J., Arabie, P., and Meulman J. (1997c). Hierarchical clustering and the construction of (optimal) ultrametrics using L_p-norms. In Y. Dodge, Ed., *L_1-Statistical Procedures and Related Topics* (IMS [Institute of Mathematical Statistics] Lecture Notes – Monograph Series [LNMS], Volume 31). Hayward, CA: Institute of Mathematical Statistics, 457–472.

Hubert, L. J., Arabie, P., and Meulman, J. (1998). Graph-theoretic representations for proximity matrices through strongly-anti-Robinson or circular strongly-anti-Robinson matrices. *Psychometrika, 63*, 341–358.

Hubert, L, J., and Baker, F. B. (1978). Applications of combinatorial programming to data analysis: The traveling salesman and related problems. *Psychometrika, 43*, 81–91.

Hubert, L. J., and Golledge, R. G. (1981). Matrix reorganization and dynamic programming: Applications to paired comparisons and unidimensional seriation. *Psychometrika, 46*, 429–441.

Jensen, R. E. (1969). A dynamic programming algorithm for cluster analysis. *Journal of the Operations Research Society of America, 7*, 1034–1057.

Johnson, R. A., and Wichern, D. W. (1998). *Applied Multivariate Statistical Analysis (4th Edition)*. Englewood Cliffs, NJ: Prentice-Hall.

Kendall, D. G. (1971a). Abundance matrices and seriation in archaeology. *Zeitschrift für Wahrscheinlichkeitstheorie, 17*, 104–112.

Kendall, D. G. (1971b). Seriation from abundance matrices. In F. R. Hodson, D. G. Kendall, and P. Tăutu, Eds., *Mathematics in the Archaeological and Historical Sciences*. Edinburgh: Edinburgh University Press, 215–252.

Korte, B., and Oberhofer, W. (1971). Triangularizing input-output matrices and the structure of production. *European Economic Review, 2*, 493–522.

Krantz, D. H., Luce, R. D., Suppes, P., and Tversky, A. (1971). *Foundations of Measurement, Volume I*. San Diego: Academic Press.

Kruskal, J. B. (1983). An overview of sequence comparison. In D. Sankoff and J. B. Kruskal, Eds., *Time Warps, String Edits, and Macromolecules: The Theory and Practice of Sequence Comparison*. Reading, MA: Addison-Wesley, 1–44.

Landau, J., and de la Vega, F. (1971). A new seriation algorithm applied to European protohistoric anthropomorphic statuary. In F. R. Hodson, D. G. Kendall, and P. Tăutu, Eds., *Mathematics in the Archaeological and Historical Sciences*. Edinburgh: Edinburgh University Press, 255–262.

Lawler, E. L. (1964). A comment on minimum feedback arc sets. *IEEE Transactions on Circuit Theory, 11*, 296–297.

Lawler, E. L, Lenstra, J. K., Rinnooy Kan, A. H. G., and Shmoys, D. B., Eds. (1985). *The Traveling Salesman Problem*. New York: Wiley.

Lenstra, J. K. (1974). Clustering a data array and the traveling salesman problem. *Operations Research, 22*, 413–414.

MacQueen, J. (1967). Some methods for classification and analysis of multivariate observations. In L. M. Le Cam and J. Neyman, Eds., *Proceedings of the Fifth Berkeley Symposium on Mathematical Statistics and Probability, Vol. 1*. Berkeley: University of California Press, 281–297.

Marks, W. B. (1965). *Difference Spectra of the Visual Pigments in Single Goldfish Cones*. Unpublished doctoral dissertation. Baltimore: Johns Hopkins University.

McCormick, W. T., Jr., Schweitzer, P. J., and White, T. W. (1972). Problem decomposition and data reorganization by a clustering technique. *Operations Research, 20*, 993–1009.

Mirkin, B. G. (1979). *Group Choice*. (P. C. Fishburn, Ed.; Y. Oliker, Trans.). Washington, DC: V. H. Winston (original work published by Nauka, Russia, 1974).

Nemhauser, G. L., and Wolsey, L. A. (1988). *Integer and Combinatorial Optimization*. New York: Wiley.

Orth, B. (1989). On the axiomatic foundations of unfolding: with applications to political party preferences of German voters. In G. De Soete, H. Feger, and K. C. Klauer, Eds., *New Developments in Psychological Choice Modeling*. Amsterdam: North-Holland, 221–235.

Rao, M. R. (1971). Cluster analysis and mathematical programming. *Journal of the American Statistical Association, 66*, 622–626.

Reeves, C. R., Ed. (1993). *Modern Heuristic Techniques for Combinatorial Problems*. New York: Halsted Press.

Roberts, F. S. (1978). *Graph Theory and Its Application to Problems of Society*. Philadelphia: Society for Industrial and Applied Mathematics.

Robinson, W. S. (1951). A method for chronologically ordering archaeological deposits. *American Antiquity, 16*, 293–301.

Ross, B. H., and Murphy, G. L. (1999). Food for thought: Cross-classification and category organization in a complex real-world domain. *Cognitive Psychology, 38*, 495–553.

Sankoff, D., and Kruskal, J. B., Eds. (1983). *Time Warps, String Edits, and Macromolecules: The Theory and Practice of Sequence Comparison*. Reading, MA: Addison-Wesley.

Schiffman, H., and Falkenberg, P. (1968). The organization of stimuli and sensory neurons. *Physiology and Behavior*, *3*, 197–201.

Schiffman, S. S., Reynolds, M. L., and Young, F. W. (1981). *Introduction to Multidimensional Scaling*. New York: Academic Press.

Shepard, R. N., Kilpatric, D. W., and Cunningham, J. P. (1975). The internal representation of numbers. *Cognitive Psychology*, *7*, 82–138.

Sibson, R. (1971). Some thoughts on sequencing methods. In F. R. Hodson, D. G. Kendall, and P. Tăutu, Eds., *Mathematics in the Archaeological and Historical Sciences*. Edinburgh: Edinburgh University Press, 263–266.

Späth, H. (1980). *Cluster Analysis Algorithms*. Chichester, U.K.: Ellis Horwood.

Späth, H. (1991). *Mathematical Algorithms for Linear Regression*. New York: Academic Press.

Thurstone, L. L. (1959). *The Measurement of Values*. Chicago: University of Chicago Press.

Thorndike, R. L. (1950). The problem of the classification of personnel. *Psychometrika*, *15*, 215-235.

Tucker, L. R (1964). The extension of factor analysis to three-dimensional matrices. In N. Frederiksen and H. Gulliksen, Eds., *Contributions to Mathematical Psychology*. New York: Holt, Rinehart, and Winston, 109–127.

Wegman, E. J. (1990). Hyperdimensional data analysis using parallel coordinates. *Journal of the American Statistical Association*, *85*, 664–675.

Author Index

Adelson, R. M., 2, 149
Alpert, C. J., 23, 25, 149
Arabie, P., 1, 2, 4, 38, 42,
    47, 57, 64, 70, 76, 88,
    89, 149, 150, 152, 153

Baker, F. B., 75, 153
Barthélemy, J.-P., 30, 64, 149
Batagelj, V., 2, 22, 47, 149
Bellman, R., 75, 149
Bezembinder, T., 3, 149
Bock, H. H., 151
Bowman, V. J., 69, 149

Carroll, J. D., 38, 150, 151
Chandon, J.-L., 38–42, 150
Charon, I., 64, 150
Chepoi, V., 38, 150
Colantoni, C. S., 69, 149
Conrad, J., 51
Coombs, C. H., 70, 150
Critchley, F., 47, 150–152
Cunningham, J. P., 3, 155

Daws, J. T., 103, 150
Day, W. H. E., 2, 150
de la Vega, F., 47, 153
de Leeuw, J., 151, 152
De Soete, G., 38, 149–151, 154
Defays, D., 57, 132, 150
Degreef, E., 150
Delcoigne, A., 2, 80, 150
DeSarbo, W. S., 38, 150, 151
Diday, E., 47, 89, 150, 151
Dodge, Y., 153

Durand, C., 89, 151

Falkenberg, P., 4, 155
Farach, M., 38, 151
Feger, H., 149, 150, 154
Fichet, B., 38, 89, 150, 151
Fishburn, P. C., 154
Fisher, W. D., 2, 22, 23, 151
Flueck, J. A., 64, 151
Frederiksen, N., 155
Furnas, G. W., 38, 150, 151

Gelfand, A. E., 47, 151
Golledge, R. G., 2, 153
Gordon, A. D., 80, 151
Greenberg, M. G., 66, 67, 69, 88, 132, 151
Groenen, P. J. F., 57, 151
Guénoche, A., 30, 64, 149, 150
Gulliksen, H., 155
Guttman, L., 23, 151

Hall, K. M., 23, 151
Hansen, P., ix, 2, 80, 150, 151
Hartigan, J. A., 2, 22, 38, 75, 151
Heiser, W. J., 51, 72, 151, 152
Held, M., 2, 8, 75, 152
Hillier, F. S., 12, 152
Hodson, F. R., 47, 151–153, 155
Hubert, L. J., ix, 1, 2, 4, 18, 34, 38, 42,
    47, 57, 64, 69, 70, 75, 76, 88,
    89, 149, 150, 152, 153
Hudry, O., 64, 149, 150

Isaak, G., 64, 149

Jaumard, B., ix, 151

Jensen, R. E., 2, 46, 153
Johnson, R. A., ix, 153

Kahng, A. B., 23, 25, 149
Kannan, S., 38, 151
Karp, R. M., 2, 8, 75, 152
Kendall, D. G., 47, 70, 76, 151–153, 155
Kilpatric, D. W., 3, 155
Klavžar, S., 2, 149
Korenjak-Černe, S., 2, 149
Korsh, J. F., 64, 151
Korte, B., 64, 153
Krantz, D. H., 90, 153
Kruskal, J. B., 2, 82, 83, 153, 154

Landau, J., 47, 153
Lantermann, E. D., 149, 150
Laporte, G., 2, 149
Lawler, E. L., 64, 76, 78, 153, 154
Le Cam, L. M., 154
Lemaire, J., 39, 150
Lenstra, J. K., 76, 154
Lieberman, G. J., 12, 152
Luce, R. D., 90, 153

MacQueen, J., 46, 154
Marks, W. B., 4, 154
McCormick, W. T., Jr., 76, 154
McMorris, F. R., 152
Meulman, J., 42, 47, 57, 88, 89, 151–153
Mirkin, B., 89, 90, 105, 152, 154
Murphy, G. L., 92, 154

Nemhauser, G. L., 2, 154
Neyman, J., 154
Norman, J. M., 2, 149

Oberhofer, W., 64, 153
Oliker, Y., 154
Orth, B., 67, 68, 154

Pouget, J., 39, 150
Pruzansky, S., 38, 150

Rao, M. R., 2, 22, 154
Reeves, C. R., 1, 154
Reeves, N., 51
Reynolds, M. L., 4, 155
Rinnooy Kan, A. H. G., 154
Roberts, F. S., 64, 89, 149, 152, 154
Robinson, W. S., 46, 154
Ross, B. H., 92, 154
Roth, P., 51
Rzhetsky, A., 152

Sankoff, D., 82, 83, 153, 154
Schiffman, H., 4, 155
Schiffman, S. S., 4, 155
Schweitzer, P. J., 76, 154
Shepard, R. N., 2, 3, 20, 77, 155
Shmoys, D. B., 154
Sibson, R., 47, 155
Späth, H., 19, 22, 37, 75, 107, 155
Suppes, P., 90, 153

Tăutu, P., 47, 151–153, 155
Tesman, B., 64, 149
Thorndike, R. L., 14, 155
Thurstone, L. L., 3, 155
Tucker, L. R., 25, 155
Tversky, A., 153

van Acker, P., 3, 149
van Buggenhaut, J., 150
van Cutsem, B., 150
van Os, B. J., 148
von Kleist, H., 51

Warnow, T., 38, 151
Wegman, E. J., 75, 155
White, T. W., 76, 154
Wichern, D. W., ix, 153
Woirgard, F., 64, 150
Wolsey, L. A., 2, 154

Young, F. W., 4, 155

Subject Index

additive tree, 37
anti-Robinson form, 22, 46, 54, 55, 63, 67, 69, 76, 89, 98, 108

branch-and-bound, 2, 39

chain, 88
    maximal, 88
cluster analysis, ix, x, 2, 17–49
    heuristic, 92–97
combinatorial data analysis (CDA), ix, 1, 51
combinatorial optimization criteria, *see* optimization criteria
combinatorial optimization methods, *see* optimization methods
combinatorial optimization tasks, *see* optimization tasks
complete enumeration, 1, 2, 7, 11, 14
computer programs, *see* programs, computer
confirmatory CDA, ix

data array reordering, 75
data sources
    multiple, 104
DHG measure, 80
diagnostic information
    hierarchical clustering, use in, 110
    partitioning, use in, 109
    sequencing, use in, 110
diagnostics, evaluating solutions, 108
dominance data, x, 53, 62, 106

dynamic programming, ix, 2
    cluster analysis, 17–49
        heuristic, 92–97
    General Dynamic Programming Paradigm (GDPP), 7–15
    heuristic use, 90–99
    hierarchical clustering, 28–49
        constrained, 45–46
        fitting ultrametrics, 39–42
        heuristic, 95–97
        two-mode matrices, 99
    linear assignment, 8–15, 48
    ordered partitions, 83–88
    partitioning, 17–28
        admissibility restrictions, 21–25
        heuristic, 93–95
        ordered partitions, 90
        two-mode matrices, 25–28
    sequencing, 51–90
        heuristic, 97–99
        one-mode skew-symmetric matrices, 62–69
        one-mode symmetric matrices, 54–62
        path construction, 73–78
        precedence constraints, 78–83, 90
        two-mode matrices, 69–73
    unidimensional unfolding
        difficulty in coordinate estimation, 71–73

enumeration
  complete, 1, 2, 11, 14
    linear assignment, 7
  partial, 2, 39
Euclidean representation, 23, 54, 55, 69

General Dynamic Programming Paradigm (GDPP), x
graph
  additive tree, 37
  circular path, 77
  complete, 18
  complete, weighted, 30
  connected, 18
  directed, 13, 14
    acyclic, 14
    sink node, 13, 14
    source node, 13, 14
  directed path
    circular, 78, 90
    linear, 78, 90
  interval, 89
    proper, 89
  maximal spanning tree, 35
  minimum feedback arc sets, 64
  minimum spanning tree, 19, 20, 30
  ordered spanning tree, 35
  path
    linear, 74

Hasse diagram, 15
heuristic methods, x, 1, 2, 30, 35, 90–99
hierarchical clustering, 2, 28–49
  agglomerative method, 30, 35, 48, 110
  complete-link method, 93
  computer programs for, 116, 126–130
  constrained, 31, 45–46
  divisive method, 35, 48
  fitted values for, 107
  greedy heuristic, use of, 32
  heuristic, 95–97
  ordered partitions, 88
  partial partition hierarchies, 32, 36, 48
  single-link method, 19, 20, 30, 35
  two-mode matrices, 35, 99
  ultrametric, fitting, 36–42
Hungarian algorithm, 12

influential objects, 109, 110

large data sets, 90–99
lattice, 88
linear assignment, x, 2, 7–15, 48
linear programming, 12

maximal chain, 88
maximum likelihood paired comparison ranking, 64
median relation, 105, 106
multidimensional scaling, 23
multiple data sources, 104
  partitioning, 105
    ordered partitions, 106
multiple structures, 107

norm
  L_p, 37, 42, 107, 130
NP-completeness, 1
Numerical Algorithms Group (NAG), ix

optimization criteria
  max/min, 12, 14
    circular path length, 77
    path length, 75
    sequencing, use in, 53
  maximization, 8, 14
    above-diagonal sum, 64, 85, 107
    circular path length, 77
    coordinate representation, equally spaced, 59, 64, 88
    coordinate representation, skew-symmetric, 64
    coordinate representation, use of, 57, 84, 85, 97
    Defays measure, 132

DHG measure, 80, 90, 117
gradient measures, 54, 63, 84, 90
gradient measures, Greenberg form, 66
path length, 75
row or column gradient, 56, 87, 90
sequencing, use in, 53
unweighted gradient, 55, 69, 70, 89
weighted gradient, 55, 69, 70, 89
within row (column) gradient, 70, 90
min/max, 12, 14
circular path length, 77
hierarchical clustering, use in, 30
path length, 75
sequencing, use in, 53
subset heterogeneity, 18
minimization, 12, 14
circular path length, 77
path length, 74
subset heterogeneity, 18
sum of transition costs in hierarchical clustering, 29
ultrametric quadruple inconsistencies, 43
ultrametric triple inconsistencies, 43
optimization methods
branch-and-bound, 2, 39
complete enumeration, 1, 2, 11, 14
dynamic programming, 2
greedy, 30, 32, 33, 93
heuristic, 1, 2, 30, 35, 90–99
Hungarian algorithm, 12
linear programming, 12
partial enumeration, 2, 39
simplex algorithm, 12
optimization tasks
cluster analysis, 17–49
heuristic, 92–97

hierarchical clustering, 2, 28–49
constrained, 45–46
heuristic, 95–97
two-mode matrices, 35, 99
linear assignment, 2, 7–15, 48
ordered partitions, 83–88, 90
partitioning, x, 2, 17–28
admissibility restrictions, 21–25
heuristic, 93–95
two-mode matrices, 25–28
sequencing, 2, 51–90
heuristic, 97–99
one-mode skew-symmetric matrices, 62–69
one-mode symmetric matrices, 54–62
path construction, 73–78, 90
precedence constraints, 78–83
two-mode matrices, 69–73
unfolding, unidimensional, 2, 9, 46, 51, 63, 70
Greenberg form, 66–69, 88
order
simple, 88
total, 88
order inversion, 40
ordered partitions, *see* partitioning, ordered partitions
outliers, 109, 110

partial enumeration, 2, 39
partial order, 88
partially ordered set (poset), 15
partitioning, 2, 17–28
t-mode matrices, 28, 103
admissibility restrictions, 21–25
t-mode matrices, 28
circular order(s), 25, 103
multiple linear orders, 25
two-mode matrices, 26
computer programs for, 115, 118–126
fitted values for, 107
heuristic, 93–95

ordered partitions, 83–88, 90
　admissibility restrictions, 87
　two-mode matrices, 87
　two-mode matrices, 25–28
path
　circular, 77
　definition, 14
　directed
　　circular, 78, 90
　　linear, 78, 90
　length, 13
　longest, 13
　optimal, 54, 74
　　computer programs for, 117
　shortest, 13
personnel classification problem, 14
poset, 88
profile smoothing, 75
program, computer
　HPHI1U, 48
　HPHI2U, 48
programs, computer, 115–148
　availability on the World Wide Web, 118
　DPCL1R, 5, 24, 47, 48, 115, 118, 122, 124, 129
　DPCL1U, 5, 46, 47, 99, 115, 116, 118, 119, 122–125
　DPCL2R, 27, 48, 115, 118, 124
　DPCL2U, 27, 47, 99, 115, 118, 123
　DPDI1U, 48, 116, 126, 129
　DPHI1R, 45, 49, 116, 126, 129
　DPHI1U, 5, 32–34, 40, 41, 48, 49, 99, 116, 126, 129, 130
　DPOP1U, 87, 90, 99, 118, 144
　DPSE1U, 59, 61, 66, 88–90, 99, 117, 131, 134, 139
　DPSE2R, 81, 90, 117, 131, 140
　DPSE2U, 71, 81, 88, 90, 99, 117, 131, 139, 140
　DPSEPH, 90, 117, 141
　HPCL1U, 99, 116, 118, 124–126, 130, 139
　HPCL2U, 99, 116, 118, 125, 130, 139, 140, 144
　HPHI1U, 44, 99, 116, 126, 130, 139
　HPHI2U, 99, 116, 126, 130, 139, 140, 144
　HPOP1U, 99, 118, 144
　HPSE1U, 99, 117, 131, 139
　HPSE2U, 99, 117, 131, 140, 144
pyramidal representation, 89

Q-form, 70, 71, 76
quasi-order, 90

Robinson form, 46, 89

sequence comparison, 82–83
　genetic, 82
　string-correction, 82
　time-warping, 82
sequencing, x, 2, 51–90
　computer programs for, 117, 131–148
　fitted values for, 108
　heuristic, 97–99
　one-mode skew-symmetric matrices, 62–69
　one-mode symmetric matrices, 54–62
　path construction, 73–78
　ordered partitions, 83–88
　precedence constraints, 78–83, 90
　two-mode matrices, 69–73
　　column order constraint, 79
　　row and column order constraints, 80
seriation, *see* sequencing
simplex algorithm, 12
sublattice, 88
subset heterogeneity measures, 18–20, 104
　k-means criterion, 19, 22, 46
　connectivity, 19
　diameter, 19, 22
　monotonicity thereof, 31
　use in hierarchical clustering, 29

traveling salesman problem, 75, 77
triangulating input-output matrices, 64

ultrametric, x, 36–42, 54, 70
    L_p norm optimal, 37, 42, 130
    admissibility
        criteria in fitting, 39–42
    basic, 37
    least-squares optimal, 29, 31, 38–42
    variance-accounted-for, 41
unfolding, unidimensional, 2, 9, 46, 51, 63, 70
    Greenberg form, 66–69, 88
    least-squares optimal, 72

unidimensional scaling, 57
    least-squares optimal, 57–62

weak order, 90
weights
    object, 104, 111–113
    proximity, indicating missing data, 104, 111–113
World Wide Web (WWW), x, 5, 118